The Evolution of Aircraft Carriers

From Pusher Biplanes of 1910
Through the World's First
Nuclear Powered Carrier – USS Enterprise (CVN-65)

By Scott MacDonald

Edited by Monty Nebinger

Acknowledgments

The Evolution of Aircraft Carriers – From Pusher Biplanes of 1910 through the World's First Nuclear Powered Carrier USS Enterprise CVN-65 is a compellation created by Defense Lion Publications (A Division of Lion Publications). This careful study of the world's endeavors to bring airpower to sea is nothing short of amazing considering the numerous powerful men in the world's navies who loved their large battleships and big guns; who felt that aircraft did not and would not play an important role in winning battles and patrolling the world's oceans and seas.

The initial thirteen chapters of this compilation are individual essays written by Scott MacDonald in the 1960s. Mr. MacDonald was a historian working for the United States Navy while writing these marvelous essays. The three Appendices at the back of this compilation were also obtained from the United States Navy and specifically within the Naval History & Heritage Command Archives. These appendices where part and parcel of the United States historical record keeping process and were developed as an aid to the navy and its personnel and to naval historians and researchers such as ourselves.

All of the images and a majority of the captions in this compilation were researched and added while creating this work. The United States Military Archives in general are a treasure trove of information; the only trouble is finding what you are looking for within millions and millions of documents and photographs with search engines that are finicky.

This compilation is dedicated to those individuals who endeavored to believe in airpower at sea; while others fervently did not. These men, both civilians and military professionals alike did not allow the status quo and what was in fashion at the time to go unchecked. It is this Editors belief that without these forward thinking individuals and the development of the super carrier that the world might be a very different place today indeed.

Copyright Notice

Copyright © 2012 Lion Publications Inc., 22 Commerce Road, Newtown, Connecticut 06470 ISBN 978-0-9859730-5-6 No part of this compilation may be reproduced or transmitted by any form or by any means, electronic or mechanical photocopying, recording or by any information and retrieval system without permission in writing from the publisher.

TABLE OF CONTENTS

Title Page	i
Acknowledgements and Copyright Notice	ii
Table of Contents	iv
Forward	v
Chapter 1 - The Aeroplane Goes To Sea	1
Chapter 2 - Decisions Out of Jutland	17
Chapter 3 - Carriers from the Keel	37
Chapter 4 - Flattops in the War Games	51
Chapter 5 - Last of the Fleet Problems	65
Chapter 6 - The Japanese Developments	79
Chapter 7 - The Early Attack Carriers	93
Chapter 8 - Emergence of the Escort Carriers	107
Chapter 9 - CVB's: The Battle Carriers	121
Chapter 10 - The End of the 'Bokubokan' In WW II	129
Chapter 11 - The Wartime European Carriers	137
Chapter 12 - The Turbulent Post-War Years	147
Chapter 13 - CVA's Built to Meet Modern Needs	159
Appendix I	1711
Naval - Aviation Chronology in World War II	1713
1940	175
1941	180
1942	189
1943	205
1944	220
1945	237
Appendix II	249
Attack Carrier Designations and Names	249
Appendix III	261
Escort Carrier Designations and Names	261

Foreword

Since February 1962, a series of articles has appeared in Naval Aviation News under the title "Evolution of Aircraft Carriers." They measure up as an authentic, earnest attempt to chronicle a history of carriers since the mobile airfield idea was initially conceived.

Here, under these covers, are the entire contents of those many articles. Inasmuch as the original naval documents concluded in 1964; Defense Lion's Naval Historians and Editors have endeavored to compile and add life to the essays with the addition of many dozens of photographs and captions.

This compilation, based on information gathered from many official sources, provides an interesting account of how and why the carrier developed as it did. It is the story behind the perhaps better known tale of carrier operations.

It is the story of change dictated by operational necessity and by technological progress. It is also the story of how naval constructors took full advantage of technological progress, and the lessons learned of operational experience to solve the Navy's unique problem of taking aviation to sea. As a result of their efforts and the constant improvement of tactics necessary to weld sea and air power together, the aircraft carrier stands today at the forefront of Naval power, ready and able to defend the nation and to project national interests to all parts of the world.

Vice Admiral, USN
Deputy Chief of Naval Operations (Air)
February 1, 1964

Supplemented by Defense Lion Publishing, October 1, 2012

This book is the first volume of a two volume series. The second volume will be titled *The Evolution of Aircraft Carriers – From the world's first Nuclear Powered Carrier, USS Enterprise Through the 100,000+ ton Super Carrier USS Gerald R. Ford.* Look for this book to be published in 2013.

Visit our Website for further details: http://www.lionpublications.com

Chapter 1

The Evolution of Aircraft Carriers

USS Intrepid (CV-11) on maneuvers off of Guantanamo Bay, Cuba 1955 - The Intrepid is now permanently docked in the Hudson River on Manhattan Island, New York. She is a floating museum of Naval and US Aerospace history. The famous SR-71 Blackbird (3+ Mach supersonic reconnaissance aircraft) that was so vital to the U.S. during the cold war is displayed on her flight deck alongside the Space Shuttle Enterprise and a retired Concorde passenger jet.
Source: www.history.navy.mil

Chapter 1 - The Aeroplane Goes To Sea

'The striking successes of carrier warfare in the Second World War are well known. Not so well known, but equally important in its own right, is the story of the evolution of sea-air power as a dominant segment in our military establishment. The formative years began almost with the birth of the aircraft itself, for the Navy was prompt to assess the value of the newest weapon in its arsenal.' -*James V. Forrestal, SecNav, 1944-1947; SecDef, 1947-1949.*

Jules Verne, author of startling science-fiction during the last half of the 19th century, would have relished some of the sketches, plans, and ideas for "aeroplanes" that crossed the desk of Capt. W. Irving Chambers in 1910. Capt. Chambers had recently been assigned as Assistant to the Secretary's Aid for Material, and was given the collateral duty of liaison between the Navy and the swelling number of letter-writers who were eager to advance their own schemes or designs involving aviation.

Less than seven years earlier, the Wright brothers had launched their pusher biplane into a brief but impressive flight. In the intervening years, advocates of aviation fought for recognition and money.

Chapter 1

The Evolution of Aircraft Carriers

At first, the Navy's interest in aviation was skeptical, if not openly discouraging. Twelve years before Chambers entered the picture, "The Joint Army Navy Board to Examine Langley's Flying Machine" was formed at the urging of Assistant Secretary of the Navy Theodore Roosevelt. A Navy member reported favorably on it to the General Board. But the Secretary, upon the advice of another Bureau in the Department, decided "the apparatus as it is referred to pertain strictly to the land service and not to the Navy."

On at least two important occasions between then and 1910, the Navy participated in or observed the fledgling "apparatus" in flight during the 1907 Jamestown Exposition and the 1908 tests by the Wright brothers at Fort Myer, Va. But the Navy Board held to the attitude that "aeronautics" had "not yet achieved sufficient importance in its relation to naval warfare" to warrant Navy support.

It was not until 1910 that specific action was taken to alert the Navy to the potentials of aviation. In one incident, pioneer Glenn H. Curtiss successfully flew a prize-winning flight between Albany and New York. At its conclusion, he prophesied publicly:

"The battles of the future will be fought in the air. The aeroplane will decide the destiny of nations." And he added, "Encumbered as [our big war vessels] are within their turrets and military masts, they cannot launch air fighters, and without these to defend them, they would be blown apart in case of war."

The "battleship controversy" was on, puffed by publicity in a competitive press. Curtiss added weight to his argument by a series of tests in which he lobbed 15 out of 22 "bombs" into targets as large as and shaped like battleships near Hammondsport, N.Y.

There was a rumor that France was building an aircraft carrier. More to the point, a growing group of enthusiasts, the U.S. Aeronautic Reserve, asked the Navy to appoint a representative who would handle aviation matters. Since this civilian organization enjoyed semi-official status, Capt. Chambers was assigned to handle all correspondence on the subject.

Chambers' job proved far from easy. He was given no space to work in, no clerical help, no operating money, no authority, and precious little encouragement. Despite this, he later wrote to Lt. T.G. Ellyson, "I am endeavoring to start an office of aeronautics here in such a way that things will run smoothly without having them all get into one Bureau and made a mess of as was the submarine question."

In October 1910, the Navy was invited to send the corps of midshipmen to Halethorpe, Md., where an aviation meet was to be held. Instead, Chambers and two other officers were sent; for the Navy, Chambers, and Naval Aviation, it was a fortunate decision. There he met Curtiss and the Curtiss trained pilot, Eugene Ely. At that time, the Navy had neither an aircraft nor a designated pilot.

Chapter 1

The Evolution of Aircraft Carriers

In a series of startling tests, Chambers, Curtiss and Ely demonstrated that this situation must change, and soon.

Several problems nagged Chambers. There was not conclusive proof, for instance, that it was feasible to launch and land aircraft at sea. And if there was to be any future for aviation in the Navy, it had to be demonstrated that aircraft could be operated in, and were important to, the Fleet. Navy officials, military and civilian, were still apathetic about the program and gave it token and grudging cognizance when they treated it with any degree of seriousness at all.

The first test was prompted by plans of a German merchant line to launch a plane from one of its ships in order to speed up its mail service. Chambers was appalled that such an advance might be made by a foreign power when the aircraft had been, in fact, developed by this country. He obtained permission to make a similar attempt at launching from the deck of the cruiser *Birmingham*. The Wright brothers were contacted, but they demurred; Ely was eager.

A temporary wooden platform was erected on *Birmingham* at the Norfolk Navy Yard. The German line, mindful of the Navy's experiment, moved up its target date in an effort to be the first to launch, and thereafter bask in the honors of claiming a significant aeronautical first. Luck was not with them, however. An accident aboard, caused by a careless workman, forced a delay of the experiment.

Chambers' plan went ahead without a hitch. On Monday, 14 November 1910, The *Birmingham* pulled into the waters off Hampton Roads, in company with three torpedo destroyers. Aboard her were pilot Ely and his biplane. Weather was unsatisfactory; visibility was dropped by a low cloud cover and there were light showers mixed with hail.

Ely's Aircraft is loaded aboard a special platform on USS Birmingham at Norfolk Virginia USA preparing for maiden flight from a US Naval War ship at anchor, 1910. Naval brass clamber to witness the spectacle while precariously standing in a mob on the angled wooden platform. Source: www.history.navy.mil

Chapter 1

The Evolution of Aircraft Carriers

Eugene B. Ely's aircraft – a Curtiss pusher biplane aboard USS Birmingham (Scout Cruiser # 2) shortly after she was hoisted aboard by a floating crane. Preparations were made and later that afternoon Ely successfully flew the Curtiss pusher off the temporary bow takeoff platform to beat the German's to the punch as the first nation to succeed in launching an aircraft from a ship. Photograph taken at the Norfolk Navy Yard, Virginia – Nov 4, 1910. Source: www.history.navy.mil

First airplane takeoff from a warship - Eugene B. Ely takes his Curtiss pusher airplane precariously off from the deck of USS Birmingham (Scout Cruiser # 2), in Hampton Roads, Virginia, during the afternoon of 14 November 1910. USS Roe (Destroyer # 24), serving as plane guard, is visible in the background. Source: Eugene B. Ely scrapbooks (www.history.navy.mil).

Chapter 1

The Evolution of Aircraft Carriers

Ely was not discouraged. He slipped into the seat of his aircraft near three in the afternoon and signaled his handlers to let loose. The plane roared off the platform, took a dangerous dip when it left the platform, then swung into the air. During the takeoff, the skid framing and wing pontoons of his biplane struck the water, nearly aborting the flight. The prop tips were splintered and water splashed over his goggles. This brief baptism, and a steady rain, blanketed his vision and for a moment he swung dizzily in the air. Finally, he spotted the sandy beaches of Willoughby Spit and touched down, ending a 2 1/2-mile flight.

Eugene B. Ely sitting in his V-8 engine powered Curtis Pusher bi-plane posing for one last photograph before hurtling down the precariously short runway of the USS Birmingham. Ely nearly ditched the aircraft after his wheel runners and wing floats sliced through the water and the propeller splintered at its tips before sailing into the air. He looks quite calm.
Source: www.history.navy.mil

The flight was an extraordinary success, but Chambers tempered his jubilance with native conservatism. Said he: "After Ely had demonstrated his ability to leave the ship so readily, without assistance from the ship's speed, or from any special starting device, such as that formerly used by the Wright brothers, my satisfaction with the results of the experiment was increased."

Chapter 1

The Evolution of Aircraft Carriers

He admitted to pre-experiment perturbation: "The point of greatest concern in my mind, carrying out the original program, was the uncertainty of stopping the ship or changing the course in time to prevent running over the aviator in case he should land in the water.

"His demonstration, that an aeroplane of comparatively old design and moderate power can leave a ship in flight while the ship is not under way, points clearly to the conclusion that the proper place for the platform is aft. An aft-platform can be made longer, will not require a lessening of the stays of any mast and its essential supports can be so rigged as a permanent structure of a scout cruiser as to cause no inconvenience in arranging the other military essentials of the ship's design."

Capt. W. Irving Chambers
Source: www.history.navy.mil

Early Aviation pioneer Glenn H. Curtiss in 1911. Curtiss was an ingenious man who built early motorcycles, automobiles and of course his famous Curtiss Pusher Biplane that was the first aircraft to take-off and land from a floating ship for the U.S. Navy. Source: blog.hemmings.com

Chapter 1

The Evolution of Aircraft Carriers

News of the feat inspired a New York Navy Yard worker to design a light movable platform for installation above the turrets in battleships for the purpose of launching aircraft at sea. Some Navy officials were enthusiastic, but Chambers was not quite so ready for this innovation. "Recognizing the practicability of Quarterman Joiner Keithley's idea," he wrote, he could "not contemplate the use of aeroplanes from turret ships in the immediate future."

Chambers' reasoning was cautious. As a result of the Birmingham flight, he did not think it necessary to launch aircraft into the wind. He had already gone on record as supporting the placement of the platform in the aft section of the ship and saw no reason to take a different stand. The safety of pilots was another determining factor: he feared they would be run over by the ship if the plane, forced to ditch, landed forward of the carrier.

Though Ely's flight opened a few Navy eyes, it did not loosen the Navy's purse strings. Glenn Curtiss, at this time, offered to teach a naval officer the mechanics of flying, absorbing the expense himself. Chambers recommended the immediate approval of the plan and Lt. T.G. Ellyson was ordered to the Curtiss camp in San Diego, CA. A series of experiments followed, in conjunction with the pilot's training.

Chambers, immensely pleased with the Birmingham launching, and was now interested in proving it practical to land a plane aboard a naval warship. Another platform was constructed at Mare Island and permission was obtained to install it on the armored cruiser USS *Pennsylvania*. While the vessel was anchored at San Francisco on 18 January 1911, Ely launched from a shore airdrome.

"There was never a doubt in my mind that I would affect a successful landing," Ely is quoted in a March 1911 Naval Institute Proceedings article. "I knew what a Curtiss biplane could do, and I felt certain that if the weather conditions were good there would be no slip."

A simple arresting gear had been installed on the ship's platform. It consisted of 22 weighted lines stretched across the deck. On Ely's plane, a number of special hooks were fitted, designed to catch the lines as the plane made its rollout. In event the jury rigged experimental arresting gear failed, a canvas screen was fitted to the end of the platform as an emergency stop.

The landing was, of course, a complete success, and Chambers was now armed with more ammunition in his battle to prove the feasibility of employing aircraft at sea. He vowed to take every opportunity to emphasize this fact to officers in the Fleet.

Chapter 1

The Evolution of Aircraft Carriers

First Landing at sea aboard U.S. Fleet ship is made by Eugene B. Ely in 1911 aboard USS Pennsylvania. Vessel was anchored in San Francisco Bay. Note the incredibly short wooden platform in which Ely must land. Arresting gear had been installed on the ship's platform consisting of 22 lines stretched across the deck and weighted with 50 lb. sand bags Source: www.history.navy.mil

Aviator Eugene B. Ely With Captain Charles F. Pond, USN, Commanding Officer of USS Pennsylvania (Armored Cruiser # 4), shortly after landing his airplane onboard the Pennsylvania in San Francisco Bay, California, 18 January 1911. Ely's wife, Mabel, is on his right and Captain Pond's wife is to Captain Ponds' left. Notice the improvised safety harness worn around Ely's torso which appears to be fashioned from a length of sturdy hose of some sort. Source: www.history.navy.mil

Chapter 1

The Evolution of Aircraft Carriers

Just 31 days after the *Pennsylvania* landing, Curtiss taxied a seaplane from his North Island base to the same ship, then in San Diego Harbor. The plane was hoisted aboard, returned to the water, and taxied back to its base. This experiment indicated the eventual liberation of aircraft from being anchored to shore bases, a necessary advancement if the aeroplane was ever to join the Fleet.

Glenn H. Curtiss flies the first successful American seaplane. Glenn Curtiss and Henry Ford were friends who shared the love of mechanical devices and speed. Ford was such an admirer of Curtiss and his ingenuity that he purchased one of the first Curtiss seaplanes. Source: www.history.navy.mil

The Navy ordered its first aircraft the following May. SecNav George vonL Meyer had earlier supported appropriations for Naval Aviation. In a meeting of the House Naval Affairs Committee he requested and received $25,000 for aeronautics.

Chambers was against the development of the true aircraft carrier by the U.S. Navy at this time. He vehemently opposed the seaplane carrier or hangar ship concept, classifying them as "auxiliary ships." He stated, "I do not believe that we need such a vessel, even if we could get it," considering it "superfluous and inefficient."

With the hydro-aeroplane, Chambers hoped to find a method of getting a plane in the air from a fast-moving vessel without being forced to slow down the ship or stop. His solution was to devise a catapult system. Langley, the Wright brothers, and Chanute had pioneered in this field, but none of the systems developed quite met the needs of Naval Aviation.

The catapult was a challenge. Chambers proposed a device using compressed air for thrust. The first test of it was made at Annapolis, with Ellyson at the plane's controls. The experiment was a failure operationally, but Chambers learned

Chapter 1

The Evolution of Aircraft Carriers

much from it. He turned the project over to Naval Constructor H.C. Richardson who, with suggestions from Ellyson and Chambers, developed it further.

Three months later, they were ready to try again. On 12 November 1912, Ellyson launched in a hydroplane, the Curtiss A-3 Triad, from a catapult installed in a barge off Washington Navy Yard. This time, they met with success. Curtiss, who witnessed the demonstration, considered it a significant

achievement.

T. Gordon Ellyson makes the first successful catapult launch of a seaplane off of a floating dock at the Washington DC Navy Yard, November 12, 1912. The catapult system used compressed air as suggested by Chambers. The aircraft was a Curtiss A-3 Triad Hydroplane.
Source: www.history.navy.mil

The following January, aviation joined the Fleet. Chambers sent the entire aviation unit to Guantanamo Bay, Cuba, to participate in Fleet operations for the first time. During the eight-week period beginning 6 January 1913, the unit conducted scouting missions and exercises in spotting mines and submerged submarines. Under specific instructions from SecNav and Chambers, the unit, led by Lt. J.H. Towers, demonstrated the operational capabilities of the aircraft to stimulate interest in aviation among fleet personnel. More than a hundred "training" flights were made, carrying interested line officers on local hops to demonstrate the safety and maneuverability of aircraft, as well as to point out the superiority of aircraft in scouting and reconnaissance tactics.

Other nations, especially in Europe, were moving faster in the development of aviation for their navies, allocating more money than the U.S. for experiments. In the same month that Chambers was officially retired, in June 1913, the British

Chapter 1

The Evolution of Aircraft Carriers

reconfigured the cruiser *Hermes* by placing a launching platform on it and using this ship actively in maneuvers that followed. The nations vied with each other in building up their air arms; in the offing were the faint rumblings that soon would swell to a roar, eventually erupting into the outrage of war.

In April 1914, Naval Aviation went into action for the first time. A crisis developed in Mexico when a U.S. naval party was placed under arrest by Mexican police. Pilots and planes were embarked in *Birmingham* and *Mississippi*. Those in the former were dispatched to Tampico and saw no action. But Lt. Patrick N. L. Bellinger, leading the *Mississippi* detachment, continued down the coast to and conducted daily reconnaissance flights.

On 5 November 1915, RAdm, W.S. Benson, the Navy's first Chief of Naval Operations, visited the *North Carolina* and a decision was made to launch the A B-2 aircraft from a new and temporary catapult installed aboard. LCdr. H.C. Mustin, who headed the Naval Aeronautic Station at Pensacola, was also aboard. He climbed into the aircraft and a successful launch was made. Though Mustin's launching was satisfactory, obvious improvements in the system were necessary. Other pilots tested the catapult, changes were made in the unit's mechanism, and finally, the catapult was removed altogether. Later a permanent catapult was installed.

Great Britain was the undisputed leader in number and operation of aircraft from ships at this time. As the U.S. was experimenting with *North Carolina*, the Royal Navy already had five vessels from which aircraft operated. First of these were *Hermes*, a cruiser converted to carry three seaplanes. Three others, formerly used as cross-channel turbine steamers, were outfitted with hangars and partial flight decks. These were *Engadine, Empress, and Riviera*, pre-*Langley* "carriers." The fifth was a converted tanker, *Ark Royal*.

Capt. Mark L. Bristol relieved Chambers in the winter of 1913. Mindful of Great Britain's progress in carrier experiments, he shot off a memorandum to SecNav: "I desire to suggest the taking up of this question at once," he wrote, "along the line of purchasing a merchant ship and converting her into an aircraft ship, and at the same time considering the plans for a special ship of this type, developing these plans as more information is received from abroad. It is strongly recommended that the bureaus consider the question of including in the estimates for the coming year money for the purchase and fitting up of such a ship with an idea of recommending to Congress the appropriations with the provision that it become immediately available without waiting until [1 July 1916]."

The memo went through the Chief of Naval Operations who sensibly felt such a venture premature. In his endorsement, he wrote: "It appears to the Department that the more immediate need of the Aeronautic Service is to determine by experience with the USS *North Carolina*, now fitted to carry aeroplanes, the details of such service upon which the characteristics of special aircraft ships, if needed, could be used." RAdm - Benson concurred with Chambers: it was not

Chapter 1

The Evolution of Aircraft Carriers

wise to spend large sums of money on carriers when the aircraft itself had not reached an acceptable state of development. There was still much to learn.

Undeterred, Bristol asked for funds for two three-million dollar carriers in his estimates for fiscal year 1917. It was a futile try. Next, he requested permission to take the command of naval air to sea and, upon receiving it, moved aboard *North Carolina*. He retained command over the Navy's aircraft, their development, the shore establishments connected with aviation, and the shaping of the air service.

Shortly after he assumed command of *North Carolina*, Bristol sailed for Guantanamo Bay to participate in war games with the Fleet. This 1916 exercise proved the most important participation of naval aircraft in any Fleet problems to date. By end of the exercise, the four planes aboard had logged more than 3890 miles in a series of tests that proved instructive and, at the same time, emphasized the lack of equipment available and that coordination and planning left much to be desired.

USS North Carolina (Armored Cruiser # 12) while fitted with a catapult aft and extended boat cranes for handling aircraft. A black arrow (left) points to a Curtiss N-9 seaplane atop the catapult in the stern. Photograph was taken at Guantanamo Bay during exercises of the war games of 1916. Photograph was a donation of Charles R. Haberlein Jr. in 2008 to the U.S. Naval Historical Center.
Source: www.history.navy.mil

In the summer of 1916, the organization, morale, equipment and prospects of Naval Aviation reached the ebb tide mark. The status of naval air so exasperated the normally reticent Bellinger that he wrote to SecNav a detailed, realistic summation of equipment available and experiments conducted. "Aeroplanes now owned by the Navy," he noted, "are very poor excuses for whatever work may be assigned them." Viewing current catapults, he continued, they are "by

Chapter 1

The Evolution of Aircraft Carriers

no means the finished mechanism desired in some of their essential features." The letter was frequently quoted by officers in the Aviation department.

With war imminent, the Appropriations Act of 29 August 1916 helped pull Naval Aviation out of the doldrums. Granted a million dollars the year before, this Act now allotted an additional $3½ million to the development of naval aviation.

In October, Towers completed a tour in London as assistant naval attaché and reported to the Executive Committee of the General Board to inform it of European progress in aviation. He spoke glowingly of zeppelins, advocated the assignment of land planes on capital ships, and discouraged the direction of attention toward aircraft carriers.

"Aeroplane ships cannot keep up with the Fleet," he reported, echoing a widely held conviction. "If [the British] build a ship big enough and powerful enough to keep up with the Fleet, its cost is so high that they do not consider it worthwhile. They are rather giving up the idea."

Towers' recommendations weighed heavily with the Board. In its subsequent recommendations, it requested over 500 planes, in addition to kite balloons, non-rigid dirigibles, and an experimental zeppelin. No recommendation was made for the fitting out of a major ship of the line for the operation of aircraft on the scope of an aircraft carrier.

The U.S. entered WW I in April 1917. In the years prior to this, Naval Aviation concerned itself with the development of aeronautical design and a continuing series of studies was implemented to determine the adaptability of planes on ships. The war interrupted these studies. Instead, emphasis was on expansion in aircraft inventory, increase in the number of trained pilots and ground crew men, and anti-submarine warfare.

In April 1917, RAdm, W.S. Sims, heading the European naval forces, recommended to SecNav that since German U-boats were sinking tremendous tonnages, attention be directed toward acquiring large numbers of seaplanes for anti-submarine reconnaissance. He also asked for the development of seaplane carriers for small seaplanes. Going a step further, he advocated the development of vessels from which seaplanes could be launched directly from their decks.

This emphasis on ASW was a reflection of the experiences of the Allied nations. Expectations of the British were high. Sims, in answering SecNav's request for information on what Allied nations' requirements for naval air support were, revealed the British preoccupation with ASW problems. Through Sims, they requested four seaplane carriers, with a capacity of six two-seater planes, six single-seaters, and a speed of at least 18 knots. They also requested four or more seaplane tenders, 100 kite balloons with necessary manpower to operate and maintain them, "any number of trained pilots," and a good 300-hp engine.

Chapter 1

The Evolution of Aircraft Carriers

But Sims appended a note of caution to these requests. He did not advise the U.S. Navy to develop this line of aeronautics if it would interfere with the completion of anti-sub programs already in progress.

Though the British pioneered in aircraft carriers, their emphasis in WW I —and that of U.S. Naval Aviation— was on the development of seaplanes. Throughout this war, seaplanes and their tenders achieved far greater attention than any other weapon in the naval air arm arsenal.

The U.S. looked for the super seaplane, one that would be large enough to carry enough fuel aboard to make a trans-ocean hop feasible. This was an attempt to circumvent the worrisome number of sinkings of cargo ships by German U-boats; with the stricken ships went a large number of aircraft built for flight against the enemy in Europe. This plane was given the designation NC and was later to prove such a flight possible.

In the summer of 1918, the General Board showed considerable interest in the future of aircraft carriers. It called before it most of the leading Naval Aviators of the day in an effort to determine how much importance to attach to this development. Testimonies presented offered a wide range of thought on the subject. Several wanted carriers for ASW work. Towers suggested the conversion of a merchant ship—for experimental purposes. Others pointed out that aircraft aboard *Huntington* were smashed by concussion when that ship fired a practice salvo. Only a ship with the major mission of launching and landing aircraft at sea would do.

The Board deliberated and in September recommended a six-year program of expansion in all branches of the fleet. For Naval Aviation, it recommended that six carriers be built within that time span, each having a 700-foot flight deck, with an 80-foot beam "absolutely clear of obstructions." Designed top speed was to be 35 knots, with a cruising range of 10,000 miles.

The bright future darkened swiftly on 2 October when SecNav Josephus Daniels temporarily put an end to the project. "The question of building aircraft carriers of special construction is held in abeyance," he wrote, "and no action will be taken until the military characteristics considered advisable by the General Board are submitted, and no action will then be taken of a positive character unless it appears probable that these vessels can be completed and made serviceable during the present war." This did not put a period to the program, simply a series of suspension dots until the Armistice.

The British had been mulling over the problem of ASW and in October 1918 proposed a possible solution to it. The proposal, at the same time, gave a keen revelation of the effectiveness of its carrier operations. Since most submarine sightings and sinkings (there were few of the latter) made by aircraft were from shore-based seaplanes, the Royal Navy suggested planes be given a much wider range than they enjoyed. They proposed a plan to tow the planes on lighters or barges to within striking distance of the targets selected. A rear compartment in

Chapter 1

The Evolution of Aircraft Carriers

the barge would be flooded sufficiently to float the plane. The aircraft would then take off, bomb its target and return to home base.

Surprisingly, the plan met with favor. The British volunteered to contribute 50 of the lighter units and asked the U.S. to provide 30, along with 40 planes. By the end of July 1918, the towed-lighter project saw the commissioning of a base at Killingholme, Ireland, with an American detachment in command. In a dress rehearsal for the scheduled bombardment of the submarine base at Heligoland, a German zeppelin appeared on the scene and photographed the entire operation. The secret type of attack no longer secret, the British called off the campaign in August.

The first draft for Naval Aviation's request for appropriations after the war contained no provision for the construction of aircraft carriers nor the conversion of a current ship of the line to carrier characteristics. But on return from Europe of Capt. Noble E. Irwin, who then had the aviation desk in the Office of the Chief of Naval Operations, the entire budget was revamped, new estimates were made, and the Navy was subsequently authorized to convert the collier USS *Jupiter* into the first experimental carrier.

The British, at that time, had three operating carriers, two training carriers and two under construction. In 1919, the General Board met again, this time centering its attention on Naval Aviation. It was an exhaustive inquiry from which was produced a report on *"Future Policy Governing Development of Air Service for the United States Navy."* In it the Board stated, **"The development of Fleet Aviation is of paramount importance and must be undertaken immediately if the United States is to take its proper place as a naval power."**

At the close of the war, the evolution of thought on carrier designs centered on the development of two types, one a fast vessel with large radius for scouting operations with scout cruisers, and the other a larger, slower vessel to operate with battleship units as a base for launching torpedo plane attacks.

The experiments and experiences of the British Navy in operating aircraft carriers influenced American thinking when design and performance were considered. Their carrier *Argus* weighed 18,000 tons and flew 20 Sopwith planes carrying 1000-lb. torpedoes. Its speed was 21 knots. Two other British carriers, *Furious* and *Vindictive*, were designed for scouting missions, traveled at 32 knots, and carried reconnaissance planes.

Arguments continued during the Board meetings. One faction wanted to convert battleships instead of colliers, but they were out-argued by Irwin who pointed out the lack of stowage space below decks, the smoke menace amidships, the small headroom between decks, and the additional personnel needed for the fire room. One admiral protested the conversion. "I believe the development is going to be so rapid that by the time you get your carriers you will find you have to make all your ships carriers." But another voice was heard, that of LCdr. E.O.

Chapter 1

The Evolution of Aircraft Carriers

McDonnell: **"A plane carrier would carry 15 torpedo planes and, in my opinion, would be a menace to a whole division of battleships and in the same way a fleet of carriers could attack a place like Hawaii."**

Congress considered converting cruisers. Merchant ship possibilities were renewed, but the Board prevailed; the collier *Jupiter* was selected.

Even at this late date, a new threat developed. After Congress authorized the carrier, RAdm, Benson shelved the project. Capt. Thomas T. Craven, who had by then relieved Irwin, found himself in the awkward position of facing a Congressional hearing and admitting that the appropriated money would not be used. He consulted Daniels who at once reversed the CNO's decision and ordered work to proceed immediately. In January 1920, Daniels allocated $500,000 for the conversion and the future of *Jupiter*. *Langley* was assured.

Several years later in 1922, LCdr. B.G. Leighton commented on the controversy surrounding the selection of *Jupiter* for the first conversion to a carrier design. "There is no good reason," he said, "why a battleship might not become an aircraft carrier, or an aircraft carrier a cruiser."

USS Langley at anchor with an Aeromarine 39-B biplane landing on her full length un-obstructed and flat flight deck in 1922. The USS Langley was originally USS Jupiter (collier#3); converted from 1920 – 1922 and served the United States Navy until 1942. Source: history.navy.mil

"The *Langley*, 14 knots, no guns, 400 officers and men—a converted collier—is an aircraft carrier. The *Saratoga*, 33 knots, eight-inch guns, three times the size of the Langley with three times as many men—a converted battle cruiser—is an aircraft carrier. The British *Argus*— a converted passenger ship is an aircraft carrier. 'Aircraft carrier' may 'mean almost anything!"

Chapter 2

The Evolution of Aircraft Carriers

Early British carrier HMS Argus in British waters, circa late 1918; seen here painted in wartime "dazzle" camouflage. Source: www.history.navy.mil

Chapter 2 - Decisions Out of Jutland

'It is impossible to resist the admiral's claim that he must have complete control of, and confidence in, the aircraft of the battle fleet, whether used in reconnaissance, gun-fire or air attack on a hostile fleet. These are his very eyes. Therefore the Admiralty view must prevail in all that is required to secure this result.'—Winston S. Churchill.

Though these words were written in 1936 as a private citizen, Winston Churchill earlier, as First Lord of the Admiralty, advocated the development of aviation in the navy while the aeroplane was still young. He was partially responsible for placing the new machines aboard British ships shortly after the first decade of this century. As a result, during World War I Great Britain developed the aircraft carrier and built a small number of them before any other country had a single ship designed for the operation of planes at sea.

Heavier-than-air craft had its start in Great Britain four-and-a-half years after Orville Wright launched the world's first successful aircraft at Kitty Hawk. Mr. Alliott Verdon-Roe completed constructing his plane at Broadside, England. Modeled after a Wright brothers' aeroplane, it was successfully flown on 8 June 1908.

On 2 March 1911, three Royal Navy officers and one Marine officer began taking flying instruction given by a civilian enthusiast. The first of the four to

Chapter 2

The Evolution of Aircraft Carriers

solo was Lt. Charles R. Samson who, in the next ten years, built a distinguished reputation for being a flamboyant man of action.

In 1912, Horace Short produced Britain's first seaplane (Churchill has been credited with coining this one word description of the aircraft) and it was successfully flown by Samson. Only months earlier, Samson demonstrated the potentials of naval aviation when in December 1911, he test launched a Shorts 27 biplane from rail platforms on the foredeck of HMS *Africa* while the warship was at anchor at Chatham. He made a safe landing alongside, using flotation bags strapped to the wheels of his plane.

The British pioneer of naval aviation, Commander Charles R Samson of No 3 Squadron, Royal Naval Air Service, standing in the forefront in full dress uniform. Note the pith helmet in his hand and the India style tents in the background; in those days all British officers knew how to live in comfort and as gentlemen regardless of their environment.

HMS Africa- at Sheerness harbor in 1912 with a Shorts S.27 pusher biplane on her foredeck rail platform. Lt Charles Rumney Samson, RN, was the first naval pilot to take off from this battleship. Samson was awarded Royal Aero Club certificate No. 71 at Eastchurch the year before.
Source: Kent History Forum

Chapter 2

The Evolution of Aircraft Carriers

Four months later, in May 1912, the first British flight from a moving ship was affected when Lt. R. Gregory, one of the "original four," took off from a temporary flight deck of the battleship *Hibernia*. The ship was steaming in Weymouth Bay at a speed of 10 to 12 knots.

By this time, France already had an Air Corps, consisting mostly of landplanes. Between 1912 and 1914, she experimented with seaplanes aboard the converted cruiser *Foudre*, previously used as a mine ship, but apparently lost interest before any notable advancement could be made. The ship could not house an effective number of aircraft aboard; the rest were hangared on the beach at Frejus. But in number of land-based aircraft in the military inventory, and in pilots trained; France was the undisputed leader in pre-WW I years.

Germany believed her future lay in the development of lighter-than-air craft, eschewing experiments in sending heavier-than-air craft to sea. Her answer to war at sea was the U-boat, supplementing the High Seas Fleet, and she used it effectively in the turbulent years ahead. She did develop landplanes, some with extraordinary achievement, but it was with Count Ferdinand von Zeppelin and his airship designs that Germany placed her national trust.

HMS Hibernia with Shorts biplane on rails. Source: Maritime Quest

Italy, at that time (and for many years after), did not believe carriers were necessary for her defense. The prevailing opinion was that the country was so centrally located it was virtually a land base from which the Mediterranean could be controlled. Japan developed aircraft carrier designs, but details of construction were not revealed to the rest of the world for decades. The United States, after originally inventing the aeroplane, did not during WW I aggressively push their operation at sea.

Chapter 2

The Evolution of Aircraft Carriers

HMS Ben-My-Chree was one of Great Britain's original seaplane carriers. This commemorative stamp shows the shorts 184 seaplane with double ailerons flying overhead near the Isle of Man 1915
Source: www.simplonpc.co.uk

A Sopwith 1 1/2 Strutter launches from the battle cruiser HMS Australia in 1918. Note the main battery gun being used as one of the braces for the extended platform required for the Sopwith to gain enough speed to take off without dipping dangerously down and close to the sea.
Source: www.history.navy.mil

True, the Navy had equipped at least three ships to operate aircraft by installing catapults on them, but the catapults were removed during the war. On the whole,

Chapter 2

The Evolution of Aircraft Carriers

the military was not encouraged and seldom financed; civilians had little motivation for building carriers.

With France the undisputed master of the landplane, Germany the acknowledged expert in lighter-than-air craft, and the whole of Europe feeling the faint stirrings of unrest as early as 1912, Great Britain was intent on catching up with and overtaking, if possible, France and Germany in their respective aeronautic specialties.

As war years approached and the German submarine force grew in potential, Britain, as the major sea power, instinctively sought ways of adapting aeroplanes for operations with the fleet while out of flying range from home bases. Her success eventually gave her a weapon more powerful than those developed by competing powers.

The genesis of the British aircraft carrier can be plotted with simplicity. At first, attention was directed to the launching of aircraft from water. Both hydroplanes and flying boats were studied, tested, and developed.

Later, experiments were made in launching planes from ships, followed almost immediately with efforts to successfully retrieve them at sea.

Eventually, the performing advantages of the light landplanes over the awkward hydroplanes led to efforts to develop vessels which could take the landplane to sea. When these achieved success, the forerunner of modern aircraft carriers was born. The gestation period was surprisingly short for such a complicated ship, but its parturition was forced by the pressures of wartime and an instinctive fight for survival.

The first group of cadets at the Central Flying School at Upavon Downs - In fact this is the very first class of cadets at the school. The commandant, Capt. Godfrey Paine RN, is seated in the front row, at the center. Major Hugh Trenchard is standing in the second row, shown at the extreme right. Below Trenchard, Lieutenant Frank Kirby VC is shown seated, also at the extreme right. Apparently one mascot was insufficient for such a group, so the photograph depicts three mascots with the cadets. Source: Air Publication 3003. HMSO – Copyright free

Chapter 2

The Evolution of Aircraft Carriers

Britain's first step toward carrying aeroplanes to sea was to establish an official air arm. On 13 April 1912, the Royal Flying Corps was constituted by Royal Warrant and, on 19 June, a Central Flying School was opened at Upavon Downs. Both the Corps and the School were planned for the centralization of aviation activities in the Royal Navy and the "Military."

Between 1912 and the outbreak of hostilities in August 1914, Europe became increasingly restless. In October 1912, following the establishment of the Corps, Britain commissioned a number of naval air stations for coast guard duty. One was placed at Cromarty, Scotland, and the remaining three in England, by the Channel coast at Calshot, Yarmouth, and Felixstowe. Two others were already in operation, one at Eastchurch and the other on the Isle of Grain. The sites were selected to form a chain so that planes could fly from one station to the next without requiring an inter-stop for refueling.

British naval aviation moved more closely toward the carrier concept when a wheeled launching platform was installed in the cruiser *Hermes* in June 1913. At first, two seaplanes operated from the ship. Later, she was capable of carrying a third. By October 1914, *Hermes* had been fitted to handle ten.

Winston Churchill, First Lord of the Admiralty. Source: British Royal Navy Photo Archives

In the summer of 1914, Winston Churchill was appointed First Lord of the Admiralty, comparable to the Secretary of the Navy in the U.S. In a series of sudden decisions, Churchill immediately called out of retirement brilliant Lord Fisher, a cantankerous admiral who advocated great changes in the Royal Navy. He was made First Sea Lord (i.e., CNO). Almost at the same time, Churchill elevated the bellicose Sir John Jellicoe to command the Home Fleet, bypassing several senior officers en route.

Aviation fascinated Churchill. He flew at every opportunity and encouraged the development of aircraft for the Navy's use.
In this respect, he was militant. In the words of Sir Sefton Brancker, then Deputy of Military Aeronautics, "The first sign of Churchill's policy was his sudden announcement that the Naval Wing of the Royal Flying Corps had become the Royal Naval Air Service—this without any reason or warning to the War Office."

His most startling decision was made shortly before war was declared. On his own initiative, Churchill called up full mobilization of the Navy, risking a veto by the Cabinet and not waiting for a signature from King George V. The entire reserve strength went on active duty; the ranks of naval aviation broadened with

Chapter 2

The Evolution of Aircraft Carriers

other units of the fleet. It was one of the few times in history that a defending nation's navy was adequately prepared upon the declaration of war.

Events moved swiftly. On 28 June 1914, the Austrian Archduke, Franz Ferdinand, was assassinated by Serbian students at Sarajevo. On 17 July Churchill concentrated the fleet at Spithead for review and maneuvers. All available naval aircraft took to the air: 17 seaplanes and two flights of aeroplanes. On 28 July Austria and Hungary declared war on Serbia. Russia sided with the Serbs and Germany mobilized. On 1 August, the British planes at Eastchurch were tuned up. August 4th, England declared war on Germany, and Germany declared war on Belgium.

HMS Hermes Sinking 30 October 1914
Source: British Royal Navy Photo Archives

At that time, Great Britain had only one vessel that could even remotely be referred to as an aircraft carrier, HMS *Hermes*. Her wartime activity was cut short, however. On the evening of 30 October 1914, she was torpedoed and sunk. Fortunately, most of her crew survived.

In short order, an old merchantman was placed in a shipyard and her superstructure converted to carry and launch seaplanes from wheeled trolleys. It was the same type of installation used in the Hermes. The merchantman displaced 7,450 tons, was slightly longer than 350 feet, and had a speed of about 11 knots. This ship, HMS *Ark Royal*, was to prove valuable to the Royal Navy in future years.

In quick succession, other vessels were converted. The former fast cross-channel packers, *Empress*, *Engadine*, and *Riviera*, were fitted with hangars for seaplanes and equipped with cranes for hoisting aircraft into and out of water. Later, an Isle of Man packet, *Ben-my-Chree*, was refitted for seaplane operations.

Except for submarine activities— which proved deadly in the early years of the war, the German Navy seemed tenaciously timid. The Kaiser adamantly refused to permit the High Seas Fleet to engage the British, so it hung reluctantly to safe ports. There were, therefore, few demonstrations of German belligerence by surface ships at sea. But in the early months, two engagements are notable, for they eventually affected some future designs of Royal Navy ships. In September 1914, the German cruiser *Konigsberg*, skulking in the Indian Ocean, attacked and sank the British cruiser *Pegasus* in port at Zanzibar. She then hid in a maze of channels in the Rufiji Delta on the east coast of Africa. The Admiralty knew her whereabouts, but not exact location. Charts indicated five possible exits for *Konigsberg*, but there was only one ship in the area able to offer chase; *Kinfauns Castle*.

Chapter 2

The Evolution of Aircraft Carriers

In September 1914, the German cruiser *Konigsberg*, skulking in the Indian Ocean, attacked and sank the British cruiser *Pegasus* in port at Zanzibar. She then hid in a maze of channels in the Rufiji Delta on the east coast of Africa. The Admiralty knew her whereabouts, but not exact location. Charts indicated five possible exits for *Konigsberg*, but there was only one ship in the area able to offer chase; *Kinfauns Castle*.

Light cruiser SMS Konigsberg – One of Kaiser Wilhelm II's war ships for the Kaiserliche Marine (German Navy) on station in the Indian Ocean in 1914. Two of her three boilers are burning as she cruises off the coast of India Source: Suite101.com – Photograph is copyright free

Not far away, on the island of Niororo, a civilian stunt pilot, H.D. Cutler, suddenly found himself commissioned in the Royal Naval Air Service and his two weathered Curtiss flying boats in the Air Service's inventory. He was immediately assigned to locate the cruiser. Only those familiar with the vagaries of war can appreciate the actions that followed.

Curtiss Flying Boat seen here flying in San Diego 1915 – Source: www.afhra.af.mil

Chapter 2

The Evolution of Aircraft Carriers

On his first flight, Cutler had no compass, got lost, was forced to beach on a deserted island and awaited rescue. *Kinfauns Castle* found him. Two days later, his leaky boat repaired, he found the German cruiser deep up a tideway. He returned to the ship and reported. Charts at the home office indicated the water too shallow to support a ship of the *Konigsberg* draft; another recon was ordered by the Admiralty, this time with an observer aboard.

RMS Kinfauns Castle; built in 1899 and transferred to the service
Source: British Royal Navy Photo Archives

Ten days were lost while Cutler awaited shipment of his second Curtiss; the first now leaked so badly it was unusable. The ship's commanding officer observed during the next flight and confirmed the *Konigsberg*'s location.

Sinking of the German cruiser now became an idee fixe with the Admiralty. The nearest ship of sufficient size and firepower to do the job was too far away. Days passed, while *Kinfauns Castle* awaited help. Cutler launched again to ascertain *Konigsberg*'s continued presence, but shortly after reaching the tideway, his engine failed. Forced down, he was captured by the Germans. Aerial reconnaissance no longer a threat, *Konigsberg* saw no reason for leaving her safe anchorage.

It was not until April that Shorts seaplanes arrived on the scene to take up Cutler's recon missions. One of the planes was shot down on its initial flight before completing a photo run. Use of the others was limited: they could not reach sufficient altitude for bombing.

Chapter 2

The Evolution of Aircraft Carriers

Shorts Seaplane experiments with the first torpedo dropped from a seaplane. This Central Flying School aircraft is dropping a 14" torpedo as the nose is pulled up.
Source: Central Flying School Photo Archives

Two more months went by before help finally came—in the monitors *Severn* and *Mersey*. They were equipped with Henri Farmans for spotting, but even then their job was not easy. A spirited fight ensued between the ships, interrupted by a five-day interim for necessary repairs to the Farmans. The battle then resumed and eventually, under persistent British gunfire directed effectively by the aircraft, the German cruiser sank.

The third German-British naval engagement of WW I has been entered in history books as the **Battle of the Falkland Islands.**

Over on the China Station, Germany had eight cruisers operating in these and other nearby waters. When Japan declared war against the Central Powers, the German squadron, commanded by Adm. Count Von Spee, sailed for South America, bombarding Papeete and Fanning Island en route. He was joined by two more cruisers at Easter Island and, in company; they proceeded to the coast of Chile. The Admiralty, intent on destroying this enemy force, assembled as many ships as possible off the southeast coast of South America, and even dispatched three from the Grand Fleet to join in the hunt.

Von Spee, still eager for battle, decided to attack the Falkland Islands. It was a fatal decision: the British Squadron came upon him unexpectedly and sank all the German ships, save one, which managed to escape.

Chapter 2

The Evolution of Aircraft Carriers

The Felixstowe F-3, called "Large America," was a British improvement of the Curtiss flying boat built before the United States entered World War I; her hull is made of wood for weight reduction. Source: British Royal Navy Photo Archives

Sopwith Camel shows machine guns installed on top of the engine cowling. Synchronizer developed by Anthony Fokker permitted the machine guns to fire precisely in between the propeller blades as it was rotating at full power. Source: RAF Photo Archives

These two incidents—the spotting and sinking of the *Konigsberg* and the Battle of the Falkland Islands—led to the later development of gun-turret launching experiments in HMS *Repulse*, and the construction of Lord Fisher's "Hush! Hush!" ships, *Courageous*, *Glorious*, and *Furious*.

The British turret-launching system was designed and developed in 1917. By early 1918, nine battle cruisers and two light cruisers were equipped to launch seaplanes from systems installed over ships' gun turrets.

Though developed by the British under the pressures of wartime urgency, the idea was first recorded as early as November 1910 when New York Navy Yard quartermaster joiner E.C. Keithley proposed a design shortly after Ely's

Chapter 2

The Evolution of Aircraft Carriers

successful take-off from the Birmingham. Keithley's idea was rejected—too advanced for its time, tossed into Navy files and forgotten.

But Fisher's "Hush! Hush!" ships have fascinated naval architects and historians since they were uncovered. Originally, they were built as cruisers of a sort under the war emergency program.

"Ships of the Royal Navy" describes them as white elephants. "In design," it states, "they suffer from being too strong and too weak. For light cruiser work, they are ludicrously over-gunned, while the absence of armor precludes their being employed as battle cruisers."

Apparently, the First Sea Lord wanted powerfully armed ships of high speed, capable of navigating very shallow waters. Officially described as light cruisers, they were ordered shortly after the sinking of *Konigsberg*. Subsequently, all three were converted into carriers, *Courageous* and *Glorious* after the war. Before *Furious* was commissioned in July 1917, she underwent the first of several conversions and emerged from the shipyard initially as an awkward-looking aircraft carrier.

HMS Furious was converted from a light battle cruiser to one of Great Britain's first serious aircraft carriers. The HMS Furious is shown here from above with a formation of Blackburn Baffin Torpedo Bombers flying off her starboard quarter. Source: British Royal Navy Photo Archives

Chapter 2

The Evolution of Aircraft Carriers

Britain, in the first months of the war, realized the danger of zeppelin raids on home shores when the Germans became entrenched in Belgium. A series of air patrols in the Channel was immediately established, costing the Royal Naval Air Service in casualties of a number of seaplanes and pilots.

*WW I Avro 504 series biplanes were used extensively by the Royal Naval Air Service.
Source: British Royal Navy Photo Archives*

In December 1914, the British planned a raid on zeppelin bases at Cuxhaven. This time, they tried a new tactic, launching the attack with seaplanes based aboard ships. The converted *Engadine*, *Riviera*, and *Empress* were pressed into service, accompanied by a screen of destroyers and submarines. The mission was not restricted to the bombing of the airship sheds, but broadened to obtain as much information as possible on the strength of the German Navy in the area.

On Christmas morning, the ships converged at a point some 12 miles north of Heligoland. An hour later, seven planes took off. En route, they were attacked ineffectively by two zeppelins, and, as they neared the enemy's main naval base, by seaplanes.

Three hours after launching, three of the seaplanes returned to their ships, the mission only partly accomplished. The remaining four were forced to ditch. The crews of three were rescued by a friendly submarine; the fourth was captured by a Dutch trawler.

The seaplanes did not succeed in finding the zeppelin sheds, thus failing that aspect of the mission. But they did bring back valuable information on harbors and the number of German ships in them. The Admiralty was not disappointed.

If any single action gave birth to the concept of aircraft carrier operations, says one noted U.S. naval historian, this raid would qualify. Several similar raids were made in later years of the war, but attention was directed first at the development of seaplanes and then of flying boats. It was not until the last

Chapter 2

The Evolution of Aircraft Carriers

months of the war that Britain fully realized the limitations of seaplane characteristics and the superiority of landplanes. She then began various experiments with true aircraft carrier design.

Meanwhile, Turkey refused to remain neutral. Influenced by Enver Pasha, the Minister of War, the country was pro-German. On 29 October 1914, Turkish warships, in company with two German cruisers, opened fire on Odessa, Theodosia and Sevastopol on the coast of the Russian Black Sea. Russia declared war on 2 November, and England and France followed three days later. The Ottoman Front was opened.

Churchill soon conceived a brilliant strategy. Had it been successfully carried out, the war could easily have been ended in 1915. Instead, the campaign ended disastrously, and the war dragged on bloodily until November 1918.

He proposed to concentrate British Forces in the Dardanelles, defeat Turkey, and force the Germans and Austrians to deploy troops and machines to that area. The Balkan states would probably join the Allies. And Russia would make a devastating victory in the east; the Central Powers would crumble. It nearly worked.

Though opposed at home and in France, Churchill ordered the Navy into action. As soon as a force of ships was gathered, including *Ark Royal*, the British armada headed toward the Dardanelles to force an entrance.

In *Ark Royal* were six two-seater seaplanes and two single-seater landplanes. Of these, only a Shorts seaplane, equipped with a good engine, was efficient. The rest could barely get high enough for effective spotting and could launch only when waters were calm.

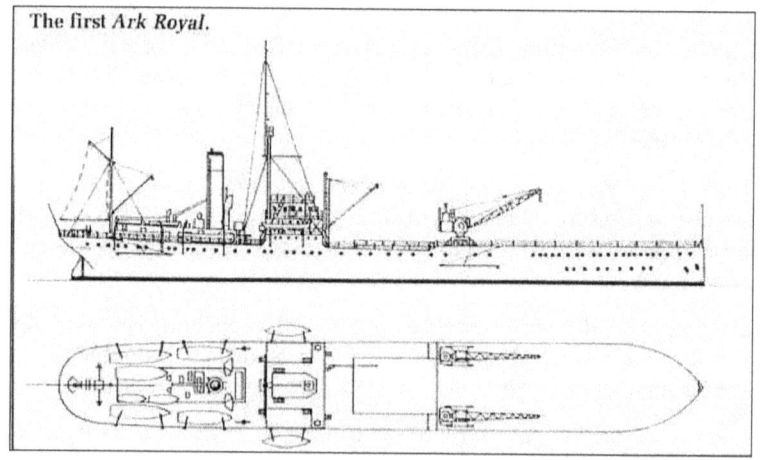

The first *Ark Royal*.

HMS Ark Royal was the Royal Navy's first Aircraft Carrier. This sketch shows the cranes in the bow to hoist seaplanes over her side to launch and recover them. Source: en.valka.cz

Chapter 2

The Evolution of Aircraft Carriers

On 5 March 1915, a Sopwith seaplane, manned by a pilot and observer, took to the air. The plane was to direct fire on a Turkish fort for the guns of the new superdreadnought, *Queen Elizabeth*. It climbed torturously to 3000 feet and, as the observer readied to call the shots, the propeller fell off. The Sopwith plunged to the sea, under furious fire from the fort. Miraculously, both men were saved.

More catastrophes followed. The assault force, entering the straits, ran into a mine field and lost three battleships. Action was broken off abruptly by the admiral—although other ships had managed to toss the Turkish and German troops into confusion.

Churchill composed a telegram insisting the battle be resumed immediately, but was dissuaded by the Admiralty on the ground that the officer commanding the situation should be allowed to make his own decisions. For the prospect of a shortened war, later events proved this decision was unfortunate.

At war's end, German General Liman von Sanders, in charge of the Dardanelles during the battle, wrote, "If the orders given at that moment had been carried out, the course of the war would have been changed after the spring of 1915, and Germany and Austria would have been constrained to continue the fight alone."

The attack on the Ottoman Front next centered on Gallipoli, but this proved a worse disaster. The enemy learned of the next tactic and buttressed their defenses. The campaign—doomed to drag on till the following January—was lost.

Samson arrived on the scene, via brisk battles at Dunkirk and Belgium, commanding No. 3 Aeroplane Squadron. *Ark Royal* moved to the Gulfs of Enos, Smyrna and Xeros, providing effective spotting, and returned to her base at Mudros. Fighting was sporadic, both a success and a failure—in about equal measure. The Turks were worthy adversaries.

By late June the threat of German submarines in these waters was real, and *Ark Royal* was retired to the safety of Imbros where she functioned as a depot ship. Barely a week earlier, *Ben-my-Chree* was added to the force. Reconnaissance and spotting flights were frequent, but the Dardanelles campaign was now at a stalemate.

In early August, a major landing was affected by the British at night without opposition. With the enemy forces nearly all routed and running, the general in charge failed to press the attack, In the meantime, reinforcements came up and the battle raged anew, continuing until the British realized the hopelessness of the situation and evacuated, ending the campaign.

Great Britain recognized the deadliness of the German U-boats early in the war. *Lusitania* was torpedoed 7 May 1915 with 1200 lives lost; 139 Americans were among them. Britain searched for a long-range seaplane that was capable of carrying heavy bomb-loads. In 1914, Sopwith developed a flying boat he called a Bat, but it was inadequate.

Chapter 2

The Evolution of Aircraft Carriers

A year later, Cdr. J.C. Porte was given command of the Felixstowe naval air station. He took up the problem, started with Curtiss flying boat designs, added improvements, and finally produced an operational craft that weighed between four-and-one-half and six-and-one-half tons. As Porte described them, they "carried sufficient petrol for work far out from land and big enough bombs to damage or destroy a submarine otherwise than by a direct hit." Called Large Americas, they were operational by the spring of 1917.

Squadron Commander Dunning's Sopwith Pup slipping over the side of the HMS Furious during his second and last (fatal) attempt to land. Men were frantically trying to save him. Source: British Royal Navy Photo Archives

Until 1915, vessels converted for aviation at sea were designed as seaplane tenders. This year, a new experiment was tried and proved successful. The Isle of Man packet, *Vindex*, was refitted to launch landplanes as well as seaplanes. A 64-foot-long deck was mounted on the ship, and a successful flight from it was made on 3 November by a Bristol Scout. The Scout seaplane was equipped with wheels which dropped off as the aircraft took to the air. It made a water landing, taxied alongside the ship, and was hoisted aboard again. Refitted with wheels and refueled, the plane was once more ready to fly.

Two other experiments were made in attempts to launch aircraft at sea to provide wider range. In the first, British Navy men designed a floating barge upon which seaplanes were towed. Nearing target, the aft compartments of the lighter were flooded, permitting the plane to slide easily into the water and take off. A variation of this was a larger platform from which small landplanes were launched. They enjoyed a brief popularity and operated in the North Sea early in the war. In the closing months of hostilities, a Sopwith Camel was launched in the same area, engaged and downed a zeppelin. The towed lighter was not refined further and saw comparatively little action.

The second experiment made by the British in 1916 tried a new approach toward launching aircraft at sea. On their own initiative, two naval officers made a design that was a departure from the standard envelope-gondola airship. The envelope they used was comparatively small, but they hoped, capable of lifting an F.E. 2C airplane. Once aloft and sufficient power given the plane, the envelope was to be detached.

Bizarre? Perhaps. At any rate, a trial launching was made of the contraption on 21 February. The plane lifted off successfully and was gaining altitude when the envelope detached prematurely. One of the officers was spilled from the plane and the other crashed with it.

Chapter 2

The Evolution of Aircraft Carriers

In Mid-1916, the war's major sea battle was fought, the Battle of Jutland. Earlier in the year, the 20,000-ton Cunarder *Campania* was converted by the British to carry seaplanes and was assigned to Adm. Jellicoe's Grand Fleet.

May approached and nearly ended before the German High Seas Fleet, now under Adm. Reinhard Scheer, made a definite move to encounter the Royal Navy. Jellicoe was ready. Advised in advance that a squadron of German battle-cruisers had been ordered to Norwegian shores for a show of force, he ordered Adm. Sir David Beatty, leading a similar but larger British squadron, to intercept.

HMS *Engadine*, operating with Beatty's squadron, launched a seaplane even though outnumbered, the German ships under Adm. Franz von Hipper, sank two of Beatty's vessels. Scheer's High Seas Fleet crested the horizon, and Beatty led his remaining ships on a strategic retreat, north toward Jellicoe.

On the day before, *Campania* had conducted a series of successful gunspotting training flights, returned to her Scapa Flow anchorage about five Jellicoe assumed his aircraft "carrier," *Campania*, was in company. Thus Jellicoe at Jutland fought without benefit of aerial observation.

Briefly, about 1800 on the 31st, the High Seas Fleet met with the Grand Fleet. Jellicoe made a thrust to cut off Scheer's retreat, but the German admiral ordered his ships first south and then east. By this maneuver, he came up in pursuit along the flank for reconnaissance at 1530 on the 31st.

HMS Eagle became Britain's second aircraft carrier. Originally planned as a Chilean dreadnought battleship, she was converted at the end of WW I. Lessons learned from construction and operations of HMS Argus were applied to the Eagle and further tests were made.
Source: British Royal Navy Photo Archives

Chapter 2

The Evolution of Aircraft Carriers

HMS Hermes the second vessel with that name in the operation of aircraft at sea, is the first aircraft carrier built as such from the keel up. As HMS Eagle learned from tests in HMS Argus, so too HMS Hermes profited from tests conducted on the HMS Eagle Source: www.roll-of-honour.com

The pilot reported three enemy cruisers and ten destroyers taking a northwesterly course. Fifteen minutes later, the German ships changed course to the south. The pilot tried to flash this signal by searchlight, but his message was not received. One of the ships of the squadron noted the alteration, however, and the ships shifted in time. Thereafter, poor visibility and rough water kept Beatty's plane on deck.

The two squadrons clashed and, miles from the main fleet, and awaited orders.

At 1735, a signal was flashed to all ships of Jellicoe's fleet to stand by to get under way. At 1900 the order to raise full steam was given and two-and-a-half hours later, *Campania* was ready. At 2254, the "proceed" signal was flashed — but the *Campania* did not receive it. Several hours passed before her C.O. realized that the rest of the fleet had gone.

Until 0200 the following morning, of the British ships, turned again and launched torpedoes, forcing Jellicoe to retreat.

Scheer then ordered Hipper to engage Jellicoe's attention while the High Seas Fleet maneuvered for an escape route. Scheer found it by 2100, cutting east across the southerly-moving British ships, and dashed to safety.

At battle's end, each fleet had lost several ships, but the British suffered more heavily in tonnage—by almost double. In post-battle retrospect, the Battle of Jutland could easily have ended in a triumphant victory for the Allies, had Jellicoe had the advantage of *Campania*'s plane to report movements of Scheer's ships. The German fleet had no seagoing aircraft. This, combined with lessons

Chapter 2

The Evolution of Aircraft Carriers

already learned in previous sea encounters with the enemy—especially in countering U-boats—strengthened more than ever the British Navy's dedication to the perfecting of the aircraft carrier.

In February 1917, the pacifism of a patient president broke when, on the last day of January, Kaiser Wilhelm notified Woodrow Wilson and the American people that unrestricted submarine warfare would be commenced on the following day. Diplomatic relations were severed on 3 February, but the President decided to wait until the next overt act before asking Congress to declare war.

He did not have long to wait. In February and March, several U.S. ships were sunk and in March, the British Secret Service obtained the famous Zimmerman note, detailing German plans against the U.S. The note was deciphered and passed on to the Americans. Wilson sent his war message to the Senate on 2 April and war was declared four days later.

Advances in British naval aviation were rapid in the closing years of the war. *Furious* joined the fleet, and experiments on landing aircraft aboard were conducted. The first attempt was successful, though unorthodox; no mechanical arresting gear was used.

On 2 August 1917, a Sopwith Pup landed aboard. On deck, handlers grasped hold of lines from the plane's wingtips as soon as the motor was cut and the plane was skidding to a stop.

In the next attempt two days later, a tire burst upon touchdown, the plane folded over the side, and the pilot was killed. Further studies were conducted and a primitive arresting arrangement was installed, along with netting to protect the ship's bridge.

Other conversions followed promptly. A cruiser of the Hawkins class was fitted with a flight deck and commissioned as HMS *Vindictive*. This deck was removed after the war.

In 1917, three ships were planned for conversion to carriers, but work was delayed intentionally on two of them. All three figured prominently in Britain's post-war development.

The first of these was the *Argus*, originally designed as the Italian liner *Conte Rosso*, and is generally considered the first true aircraft carrier. *Argus* had a flight deck 558 feet long by 60 wide and displaced 14,450 tons. She was the first "island" carrier. Her superstructure moved to a tight location on the starboard side of the ship.

The second was commissioned HMS *Eagle*, but was originally laid down as the dreadnought battleship *Almirante Cochrane* under a contract with Chile. War interrupted completion of the ship, contracts were renegotiated, and she was converted to an "island" carrier. She was the only aircraft carrier to have two funnels.

Chapter 2
The Evolution of Aircraft Carriers

HMS *Hermes*, the second carrier to bear that name, was designed from the keel up to operate as a carrier, the first such vessel constructed.

Argus was the first completed, but saw no action in the war. Convinced now that the progress of sea power lay in the future of aircraft carriers; Great Britain suspended construction on the *Eagle* and *Hermes* until tests were made on the first carrier. The lessons learned were incorporated in the *Eagle* —and this carrier was further tested. Results from experiments on both her predecessors contributed heavily to the eventual construction of the *Hermes*.

The formative, experimental years of carrier warfare drew to a close when, on 11 November 1918, hostilities ceased and the Armistice was signed. Out of the costly, bitter fight for survival a potent new ship-of-the-line developed. Great Britain pioneered in the creation of the modern aircraft carrier.

But at war's end, the U.S. had no vessel specifically built to carry aircraft to sea. Primarily, U.S. Naval Aviation launched patrol flights from shore bases. During the expansion of military forces the Navy's General Board made concrete recommendations in favor of carrier developments. After the Armistice, it listened to exhaustive testimony concerning the role of aviation in the Navy. Acting on the Board's findings, Congress authorized a small amount of money for conversion of the collier USS *Jupiter*.

When the refitting was completed, the ex-collier was renamed USS *Langley* and commissioned on 20 March 1922 at Norfolk, Va. Surrounded by modern vessels of her day. She appeared to be the strangest-looking ship to join the fleet since the Federal ironclad *Monitor* squatted heavily in the water during the Civil War. Small and gangling as she was, USS *Langley* was the first-born of a large fighting family of powerful Navy ships.

USS Langley was originally a collier, refitted to operate aircraft and recommisioned in 1921. Flying deck was 520 feet long and 65 feet wide. Hangers beneath held seaplanes and landplanes; in later years she was converted to a seaplane tender and was sunk off Java in 1942. Source: www.history.navy.mil

Chapter 3

The Evolution of Aircraft Carriers

USS Wasp (CV-7) anchored in Casco Bay, Maine in early 1942 with SB2U and F4F aircraft on her flight deck. She is shown here painted in Measure 12 (Modified) camouflage pattern. USS Wasp was built from the keel up in 1940 and was sunk by the Japanese on 15 September 1942, while operating in the Southwestern Pacific in support of forces on Guadalcanal. Source: www.history.navy.mil

Chapter 3 - Carriers from the Keel

'Such remarks as I may have to make as to the nature and extent of the air force required by the Navy will be based upon the assumption that the airplane is now a major force, and is becoming daily more efficient and its weapons more deadly, that therefore even a small, high-speed carrier alone can destroy or disable a battleship alone, that a fleet whose carriers give it command of the air over the enemy fleet can defeat the latter, that the fast carrier is the capital ship of the future. Based upon these assumptions, it is evident that our policy in regard to the Navy air force should be command of the air over the fleet of any possible enemy.'—Adm. William S. Sims, USN, October 14, 1925

Plenipotentiaries of the United States, the British Empire, France, Italy and Japan met in Washington in the early Twenties to reach an agreement on the limitation of naval armament. The treaty they signed on February 6, 1922 had a profound effect on the evolution of aircraft carriers. From the time the U.S. Navy first embarked upon a carrier-building program, it was faced with tonnage limitations established by this treaty.

The total tonnage for aircraft carriers of each of the contracting powers permitted the U.S. and Great Britain 131,000 tons each, France and Italy 60,000 tons each, and Japan 81,000 tons. Of its allotted tonnage, the United States had

Chapter 3
The Evolution of Aircraft Carriers

already consumed 66,000 in the *Lexington* and *Saratoga*. Only 69,000 tons remained for future construction. The Navy gave much thought and study to the means of best utilizing this remainder, and, in 1927, when drawing up a five-year shipbuilding program, the General Board recommended construction of a 13,800-ton carrier each year.

The program involving this plan was promptly submitted to the President who approved it on December 31, 1927. It was subsequently submitted to Congress which, by act of February 13, 1929, authorized construction of one 13,800-ton carrier. The Navy attempted in the following years to obtain authorization for construction of the visualized sister ships, but without success. Indeed, before another carrier was to be authorized, the Navy had become more interested in larger ships of about 20,000 tons.

In addition to the legal reasons which led the Navy to seek a 13,800 ton carrier, there was a body of thinking on the part of some Naval Aviators which recognized the utility of small carriers. This was evident as early as 1925 when the General Board briefly considered but rejected the conversion of 10,000-ton cruisers to light carriers.

USS Ranger (CV-4) is viewed in March 1937. Commissioned in June 1934, she was the U.S. Navy's first aircraft carrier built from the keel up. Four more were built before World War II.
Source: www.history.navy.mil

Chapter 3

The Evolution of Aircraft Carriers

*In the later years, USS Ranger saw considerable action. A Grumman F4F-4 "Wildcat" fighter taking off to attack targets ashore during the invasion of Morocco, 8 November 1942. Note: Army observation planes in the left middle distance; Loudspeakers and radar antenna on Ranger's mast.
Source: www.history.navy.mil*

Two years later, LCdr. Bruce G. Leighton, then aide to the Secretary of the Navy, prepared a study on possible uses of small carriers. In addition to protection of the battle line, he suggested their suitability for anti-submarine warfare, reconnaissance, and reduction of enemy shore bases.

At about the same time, RAdm, William A. Moffett argued that British and Japanese experience with small carriers had made it clear that such ships could keep more aircraft in operation than could an equal tonnage devoted to larger ships.

Fleet commanders, who might be expected to have had a more conservative view of the military utility of aircraft than did Moffett and Leighton, expounded concepts that provided further justification for smaller carriers.

For example, the Commander in Chief, U.S. Fleet, noted in his 1927 annual report that the Fleet was seriously handicapped by the absence of a carrier with the battle line upon which spotting planes could land. Thus, both the aviation protagonists and the surface commanders recognized the need for carriers which would perform important roles, even if they were not of a size approaching that of the giants, USS *Lexington* and USS *Saratoga*.

Chapter 3

The Evolution of Aircraft Carriers

USS Yorktown (CV-5) Anchored in Hampton Roads, Virginia, 30 October 1937. Boat booms are rigged out, with boats tied up to them. Note details of her stern, including name, structure supporting the after flight deck, and motor launch stowed athwartship on a platform between the main and flight decks. Source: www.history.navy.mil

Such considerations were in the genesis of CV-4. When it came to reducing them to detailed plans for construction of a new ship, very little had been done. Studies made in 1923 and 1924 had been concerned with island-type vessels, such as the *Lexington* and *Saratoga*, and were not directly applicable to a new design— which was to be of the flush-deck variety. In addition, the basic concept for CV-4 was embodied in the General Board recommendations of 1927 and predated the commissioning of *Lexington* and *Saratoga*. Hence, the concept could not incorporate any lessons learned during their early fleet operations.

This concept, as outlined by the General Board, included a speed of 29.4 knots, a clear flying deck, 12 five-inch anti-aircraft guns and as many machine guns as possible. On July 26, 1928, BuAer elaborated on this proposed design in a letter to Commander Aircraft Squadrons, Battle Fleet. The flight deck was to be about 86 feet by 750 feet and fitted with arresting gear. The navigating and signal-bridge were to be under the flight deck, well forward, with extensions beyond the ships side, port and starboard.

Chapter 3

The Evolution of Aircraft Carriers

USS Ranger (CV-4) anchored in Guantanamo Bay, Cuba with a full complement of aircraft onboard. Eight two-seater Vought O3U-3 Corsairs are seen on her starboard side amidships, just forward of her tower and mast. Source: www.history.navy.mil

USS Ranger (CV-4) loaded from stem to stern with Vought O3U-3 Corsairs. The U.S. Naval flight crews were masters of moving aircraft around on the flight decks of their ships. Source: www.zap16.com

As for the anti-aircraft battery, it had been reduced to eight 5-inch 25 caliber guns located two on each quarter. Anti-aircraft battery directors were to be provided, but BuAer thought that range finders should be omitted.

Secondary conning stations were to be located on the starboard side of the upper deck and combined with the aviation control station. A plotting station consisting of flag plot and aviation intelligence office was also to be provided.

Despite the fact that the general concept could not benefit from experiences of the *Lexington* and *Saratoga*, the two ships did comment on plans for the *Ranger* on the basis of such experience as they had obtained during the first year's operation.

Chapter 3

The Evolution of Aircraft Carriers

For example, they felt that both elevators and shop provisions should receive special consideration above and beyond that which had already been given. "Experience during the present concentration on both carriers has emphasized the importance of the after elevator in addition to the two now contemplated (for CV-4)," wrote *Saratoga's* commanding officer.

"There is required a great deal of re-spotting of planes in flight operations, and an after elevator will considerably expedite this process. After planes have landed on deck, it is sometimes necessary to send below a plane from the after part of the flight deck, which is now difficult with the flight deck filled with planes and the elevators forward."

Officers aboard both *Lexington* and *Saratoga* held informal conferences, the results of which were passed to BuAer. Speed was most desirable in aircraft carriers, but speed also had its drawbacks, as these officers were quick to point out to their superiors.

"The location of the A&R and general workshops aft is decidedly undesirable," BuAer informed the Bureau of Construction and Repair, "and it is strongly recommended that they be relocated further forward, if there is any possible way of doing so. Experience on CV-2 and CV-3 has shown that it is impossible to do any work requiring precision or accuracy, such as cutting a thread, when the ship is steaming at about 22 knots or more."

Early in the planning stage, BuAer encountered head-on the problem of lighting and night landings. A memorandum written for BuAer files pointed out: "The primary difficulty involved in night operations for airplane carriers is the provision of adequate illumination to enable the pilots to make safe landings and

USS Saratoga (CV-3) In Puget Sound, Washington, following overhaul, 7 September 1944. Note her hull's camouflage paint scheme and "3" on her flight deck. Source: history.navy.mil

Chapter 3

The Evolution of Aircraft Carriers

Captain Albert C. Read, USN, Commanding Officer, USS Saratoga (CV-3) in full dress uniform inspects the ship's Marines during change of command ceremonies at the end of his tour as her Commanding Officer, circa 15 March 1940. Source: www.history.navy.mil

at the same time to enable the ship to maintain darkened ship conditions that will prevent disclosure of the carrier's provision to surface craft and enemy aircraft. The technical difficulties in this project are so great that complete success can scarcely be hoped for, for several years and then not without the expenditure of much more time and effort than appears desirable at present.

"Night flying experiments were conducted on the *Langley* to determine the type of illuminating equipment for the *Saratoga* and *Lexington*. Although the number of landings made was not very great, enough information was obtained to determine upon equipment that would at least provide for a point of departure for future experiments in an effort to further solve the basic problem. No carrier night flying has been conducted since 1925." The memorandum was dated June 14, 1929.

This sparked an intensive series of experiments which caused the introduction of several lighting systems aboard various carriers. At best, most of these provided safe illumination for night landing but were less successful in maintaining darkened ship. Incandescent lights of low wattage were tested in various arrangements and intensities. Neon tubes were tried, some colored green, red, blue or amber. Of these, plain white was considered the best—but was not a solution. Even luminous paint was investigated. The problem of night deck illumination was to plague Navy for years to come. How the problem was

Chapter 3

The Evolution of Aircraft Carriers

handled in USS *Ranger* is indicated by a November 1934 report her commanding officer made to BuAer:

"In anything but bright moonlight when the ship's outline can be made out at a reasonable approach distance, it is very difficult, definitely too difficult, to get in the groove when only landing deck lights are used. Although *Ranger*'s landing deck lights extend the length of the ship and are well lined up on each side, which it was hoped would improve the difficulty described by *Saratoga* and *Lexington* pilots, the pilot is frequently too near the ship before he can find out which way to swerve. If he happens to hit the groove early, he is well fixed. If he doesn't, he sees a jumble of landing deck lights and can only guess whether to change course to right or to left."

"With ramp lights turned on in addition to the landing deck lights, there is unanimous agreement that getting in the groove is very easy. Exactly why this is true is not clear, but the string of lights across the ramp appears not only definitely to locate the end of the deck, but also to give the pilot sufficient basis for setting his course normal to the lights and up to the centerline of the deck."

"Athwartship landing deck lights at bow and stern are no use and would be hazardous if opened when planes are landing. (Confusion in getting in the groove existed whether or not these lights were opened, worse when opened.)"

Other problems were of concern to BuAer during the design stage of CV-4. Relatively minor, but illustrative of the care devoted to carrier design, was the question of paint color for interior surfaces. A flurry of correspondence between BuAer and BuC&R concerned the color of paint to use on the deck, overhead, and bulkheads of the hangar.

USS Lexington(CV-2) loaded with a full complement of aircraft. She carried scouting and reconnaissance planes, Vought O3U-3 Corsair bombers and others. She is seen here at anchor somewhere in the early 1940's. Note the booms on her port side. Source: www.goatlocker.org

Chapter 3

The Evolution of Aircraft Carriers

This was not so much a problem of habitability as it was one of weight limitation and maximum reflective power. White paint, light gray and aluminum were considered. Misinformation supplied the Bureau of Engineering caused it to advocate light gray, but BuAer objected. Tests were conducted and aluminum proved lighter and more reflective of the three paints considered.

Finally, in early December 1929, plans for CV-4 received approval. Copies were sent to the Fleet, noting that major changes could not be made in them, but that the Bureau would "be glad to have comment or suggestion with regard to minor points, should such comment appear desirable."

By February 1930, active work on the design of the 13,800-ton carrier had stopped. Shortly after British Prime Minister Mr. MacDonald visited the United States, the President gave instructions to suspend all work on this ship, pending the outcome of the then projected London conference on naval armament. Months went by, the President was consulted again, and again the Navy was told to do nothing about the ship until the treaty had been ratified.

The treaty was signed in London on April 22, 1930. Ratification of the treaty was advised by the Senate on July 21, 1930, and by the President on the following day. In the meantime, the Navy Department, Office of the Judge Advocate General, drafted an advertisement which was published when the ratification restriction was lifted. The advertisement invited bids for the construction of CV-4. The bids were opened September 3—and proved to be "bombs."

All bids submitted far exceeded the appropriation given the Navy for construction of the ship, the lowest bid (by Newport News Shipbuilding and Dry Dock Co.) exceeding the limit by an estimated $2,160,000. The four Navy Department bureaus involved in the construction plans— BuC&R, BuAer, BuOrd, BuEng— forwarded a joint memorandum to the Secretary of the Navy requesting a 60 day extension of the period before execution of the contract in order to consider necessary changes in characteristics which would permit construction of the carrier within the cost of the lowest bid.

Permission was obtained and the various departments reviewed their requirements. Panels of officer-experts in each were formed to submit recommendations. Out went consideration of an extra elevator. Out went the possibility—at this time—of moving the shops forward, as *Saratoga* and *Lexington* had suggested. Submitting its list of recommended savings, BuAer listed the elimination of catapults, smokestacks on one side, sliding doors for the hangars, securing tracks, and airplane booms and nets, and requested that necessary eliminations be made in that order.

"This bureau feels," wrote Cdr. R.K. Turner for BuAer, "that elimination or reduction in the balance of items considered, namely, arresting gear, elevators, or gasoline capacity would seriously affect the characteristics of the ship as an

Chapter 3
The Evolution of Aircraft Carriers

aircraft carrier, and, therefore, urgently recommends against any sacrifice in these items."

By October 2, the Bureaus had signed another joint letter, addressed to the General Board, listing their recommendations on how to cope with the problem of the elimination of design features. Among other things, *Ranger*'s fire control was to be simplified, ammunition storage space was to be reduced, bombing planes were to be substituted for torpedo planes (this eliminated the purchase of torpedoes), deck catapults were to go by the boards, as were plane booms and nets. Twenty percent of the flight-deck securing tracks were to be eliminated, as well as housing palisades, and the voice tube system. Finally, the arresting gear system was to be reduced. On November 1, 1930, the contract was signed by Newport News.

The launch of USS Ranger (CV-4) at Newport News Ship Building and Dry Dock Company, Newport News, Virginia took place on 25 February 1933. Note the three elevator platforms already installed. Much more work is required to fully fit out naval ships especially aircraft carriers after initial launch including the island, mast, flight deck and so on.
Source: www.history.navy.mil.

Throughout official correspondence, the 13,800-ton carrier was referred to simply as CV-4. On December 10, 1930, the Bureau of Navigation informed a long list of addressees that "The Secretary of the Navy has assigned the name *Ranger* to Aircraft Carrier No. 4, authorized by Act of Congress dated February 13, 1929. The assignment of the name *Ranger* is in accordance with the Department's policy of giving names formerly assigned to those battle cruisers scrapped by terms of the Washington Treaty."

Chapter 3

The Evolution of Aircraft Carriers

On September 26, 1931, *Ranger*'s keel was laid. Seventeen months later, the ship was launched, and on June 4, 1934, she was commissioned. Though planned originally as a 13,800 ton aircraft carrier, she exceeded this tonnage by 700 tons. Original plans also called for a severe flush deck, but, upon commissioning, she had a small island.

USS *Ranger* had eight 5-inch 25 caliber AA guns, other AA guns in gallery. She could operate 75 aircraft and had a complement of 1788, of which 162 were commissioned officers. Her aircraft consisted of four squadrons of bombers and fighters and a few amphibians. CV-4 also was equipped with a box arresting gear—a feature included in other fast carriers until early 1943.

The General Board had become convinced—even before the *Ranger* was launched—that the minimum effective size of aircraft carriers was 20,000 tons. A request for two of these heavier ships was made in the Building Program for 1934, which was issued in September 1932. In May the following year, the Board again submitted this recommendation. As a result, the Secretary of the Navy asked the President for Public Works Administration funds to build two carriers of this tonnage, in addition to other ships. USS *Yorktown* (CV-5) and USS *Enterprise* (CV-6) were authorized.

USS Yorktown (CV-5) In Hampton Roads, Virginia, with her port anchor out, 30 October 1937. Note Landing Signal Officer platform near the front of her flight deck, for use in landing planes over the bow. Source: www.history.navy.mil

Files of the Bureau of Aeronautics housed in the National Archives reveal a memorandum dated May 15, 1931, which was to affect the two new carriers:

Chapter 3

The Evolution of Aircraft Carriers

USS Enterprise (CV-6) was commissioned in May 1938, a sister ship to the Yorktown. "The Big E" was to become a popular ship. Capt. N.H. White Jr., was her first Commander. This photograph taken from USS Minneapolis (CA-36) was taken while Enterprise and others from her battle group were in route to Pearl Harbor, 8 October 1939. Source: www.history.navy.mil

"The Department has approved a new building program with two aircraft carriers similar to the *Ranger*, but before embarking on this new construction, it is suggested that a careful examination may show many design changes are desirable."

"The particular improvements in the *Ranger* design that should be considered are: speed increase to 32.5 knots; addition of underwater subdivision to resist torpedo and bomb explosions; horizontal protective deck over machinery magazines, and aircraft fuel tanks; improvement in operational facility (this includes hangar deck devoted exclusively to plane stowage, four fast elevators, complete bomb handling facilities, possible use of two flying-off decks, and improved machine gun anti-aircraft defense)."

The *Yorktown* was launched April 4, 1936, sponsored by Mrs. Franklin D. Roosevelt. When the carrier was commissioned September 30, 1937, her overall length was 827 feet, four inches; extreme beam was 95 feet, four inches; and standard displacement, 19,800 tons. Her trial speed was 33.6 knots. USS *Enterprise* (CV-6) was the seventh Navy ship to bear this name. Her keel was laid July 16, 1934 and she was launched October 3, 1936, sponsored by Mrs. Claude A. Swanson, wife of the Secretary of the Navy. She was placed in commission at Norfolk on May 12, 1938. Her specifications were similar to

Chapter 3

The Evolution of Aircraft Carriers

Yorktown's. She had accommodations for 82 ship's company officers and 1447 enlisted men.

As soon as CV-5 and CV-6 were authorized, the General Board did not request additional carriers of such tonnage. It did, however, vainly plead for a 15,200-ton replacement for the obsolete *Langley*. The *Langley* had been classed as an experimental ship and did not figure in the U.S. Navy's aircraft carrier tonnage limitations. To replace her with another carrier would have been to violate the treaty. The Navy did plan, however, to request new aircraft carriers when the *Lexington* and *Saratoga* reached retirement age.

Tightening of world tensions in 1938 caused the Navy Department to reconsider its carrier-building program, and USS *Hornet* (CV-8) was authorized on May 17 that year. She was launched December 14, 1940 and commissioned October 21, 1941, with Capt. Marc A. Mitscher, her first commanding officer.

USS *Wasp* (CV-7) had been ordered earlier, on March 27, 1934. Her keel was laid April 1, 1936; she was launched April 4, 1939, and commissioned April 25, 1940. This carrier had to be built within what was left of the 135,000-ton limit set by the treaty. She was commissioned at 14,700 tons. Thus there were left only a few hundred tons remaining of the treaty-authorized carrier strength.

Already in the mill, during construction of *Yorktown* and *Enterprise*, were plans for a new class of aircraft carrier, the first of which would be known as USS *Essex* (CV-9).

War clouds were gathering over Europe and the Pacific. Fleet exercises and war games were stepped up as international tensions mounted. The treaties of 1922 and 1930 terminated December 31, 1936 when Japan abrogated.

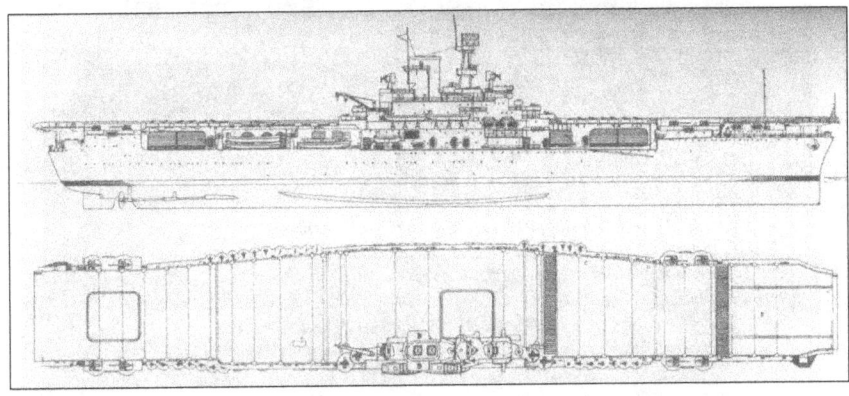

USS Wasp (CV-7) was launched in April 1939 and commissioned in April the following year. She displaced 14,700 tons, this weight restricted by the tonnage remaining from the limitations of the 1922 Naval Armament Treaty. Her first Skipper was Capt. J.W. Reeves Jr.
Source: www.history.navy.mil

Chapter 3

The Evolution of Aircraft Carriers

In its provisions for Naval Aviation, the Naval Expansion Act of May 17, 1938 authorized an increase in total tonnage of under-age naval vessels amounting to 40,000 tons for aircraft carriers, and also authorized the President to increase the number of naval aircraft to "not less than" 3000. Carriers built as a result of this authorization were the *Hornet* and *Essex*.

On September 8, 1939, President Roosevelt proclaimed the existence of a limited national emergency and directed measures for strengthening national defenses within the limits of peacetime authorization. In May 1941, an unlimited national emergency was declared. Seven months later Japanese aircraft, launched from carriers, attacked Pearl Harbor, and within 24 hours, the President went before Congress and the nation was at war.

USS Hornet (CV-8) was authorized in 1938 when world tensions mounted, launched in December 1940, commissioned in October 1941. Hornet had a standard displacement of 19,800 tons. First Commanding Officer was Captain Marc A. Mitscher. Photographed here in late 1941, soon after her completion; probably at a U.S. East Coast port; this is assumed because a ferry boat and "Eagle Boat" (PE) are in the background.
Source: www.history.navy.mil

Chapter 4

The Evolution of Aircraft Carriers

As she appeared in March 1925 when the first aircraft carrier appeared in the annual war games, USS Langley (CV-1) provided actual carrier capability that the first four games lacked. Fleet Problem V was conducted off California. It provided "valuable experience."
Source: history.navy.mil

Chapter 4 - Flattops in the War Games

'If the mind's eye is always directed upon the series of engagements, so far as it can be perceived beforehand, then it is fixed upon the direct road to its goal, and thereby the movement of our strength acquires that rapidity, that is to say, our volition and action acquire that energy which the occasion demands and which is not disturbed by extraneous influences.' —Karl von Clausewitz, On War

One of those whose untiring efforts helped shape the evolution of the "all big gun battleship," Adm. William S. Sims, did not immediately endorse Naval Aviation—especially ships carrying naval aircraft—upon its introduction as a weapon in the country's arsenal. In 1909, for instance, he wrote: "According to the papers, one of the Wright brothers has stated that it would be impracticable to hit anything by dropping a projectile from his flying [machine]. That Wright man is right, all right."

Sims had a deep appreciation and understanding of the merits of the battleship as a weapon system whose evolution he had fought to promote and he was not about to write it off, except on the basis of sound evidence. During WW I and the years immediately preceding it, aircraft design improved spectacularly. By the end of the war the U.S. Navy still did not have an aircraft carrier. His observation of the limited use of such ships permitted him to state with justification, "All the aeroplane-carrying ships in the world could not make an

Chapter 4

The Evolution of Aircraft Carriers

attack upon a foreign country unless they were supported by a battleship force that was superior to that of the enemy."

Not until the end of the war, when Adm. Sims assumed leadership of the Naval War College at Newport, did his thinking undergo a profound change. At the game board there in 1921, he recognized not only the advantages and potentials of airpower but also the brevity of the future of battleships. "If I had my way," he said, "I would arrest the building of great battleships and put money into the development of the new devices and not wait to see what other countries are doing."

By March 1922, after witnessing bombing tests off the Virginia Capes (in 1921), he had written, "The battleship is dead."

During Sims' tenure at the War College, the Navy Department inaugurated a series of war games, fleet exercises that were conducted during the next two decades. Through these problems, the Navy obtained practical experience in testing the "new devices" under simulated combat conditions.

Naval Aviation had entered fleet maneuvers as early as the winter of 1912-13 when the entire aviation element—pilots, student pilots, enlisted men and aircraft inventory, which then totaled five planes, was transported to Guantanamo Bay to take part in planned exercises. From their camp at Fisherman's Point where the present air station is located, they worked to achieve three goals: first, to prove the utility of the airplane as a scout under simulated war conditions; second, to test its usefulness in detecting mines and submerged submarines; and third, to stimulate interest in aviation among officers in the Fleet.

Naval Aviation next joined the Fleet in 1914, in connection with actual hostilities in Mexico. At that time, an A-3 and a c-3, put aboard the *Mississippi*, saw action at Vera Cruz. Daily reconnaissance flights kept landing forces informed of the enemy dispositions inshore. (Three planes placed aboard the *Birmingham* were taken to Tampico but did not see action.)

As a result of the experience at Vera Cruz, Naval Aviators judged the hydro-aeroplane more efficient than the flying boat type then in use. Recommendations were also made on the design of aircraft.

The Navy's air arm was still very small when the United States entered WW I. In the next year, seven months and four days, while war raged, its growth was extraordinary. By the time the Armistice was signed, the Navy had 2107 planes, 570 of which were overseas, 15 dirigibles, 205 kite balloons, and 10 free balloons.

Thirteen bases were established in the U.S. and the Canal Zone, only one of which, at Galveston, was not yet in operation. In Ireland, the Navy had four seaplane stations, one kite balloon station, a receiving station and a supply station. Two stations, including a major assembly and repair base, were

Chapter 4

The Evolution of Aircraft Carriers

The Back Fleet was assigned the battleships *New York* and *Oklahoma* as "constructive" carriers. Approaching the Canal, one of the battleship "carriers," the *Oklahoma*, launched a seaplane by catapult to scout ahead of the force. Early the next morning, a single plane representing an air group took off from Naranyas Cays, approached the Canal from seaward, flew over Gatun Spillway, and dropped ten miniature bombs. This plane completed its mission undetected and theoretically destroyed the Spillway.

An official report submitted after the problem pointed up the susceptibility of vital parts of the Canal to destruction by air. The report urged, among other things, that air defenses of the Canal be strengthened and that rapid completion of aircraft carriers be effected for offensive and scouting purposes.

Naval Aviation played little part in the next three exercises. It was not until Fleet Problem V in March 1925 that USS *Langley* entered exercises off the California coast. The second phase of the problems began; a new element was introduced.

This is a picture of the Lexington (CV-2), and her sister ship, USS Saratoga (CV-3) sailing in formation. The Lexington in this picture is on the far right. USS Saratoga (CV-3) and her sister ship USS Lexington entered the war games in Fleet Problem IV in 1929. The extraordinary tactic employing Saratoga in this exercise revolutionized naval strategy and spurred requests for more aircraft carriers for fleet operations. Source: wwiiguide.webs.com

Basically, the supposition for this problem was that strained relations existed between Blue (the U.S.) and Black, an imaginary country in the area of the Hawaiian Islands. When Black declared war, its Commander-in-Chief was

Chapter 4

The Evolution of Aircraft Carriers

ordered to Guadalupe Island where he was to occupy an unfortified anchorage from which he was to operate against Blue in the Eastern Pacific.

Black was given the *Langley* and the tenders *Aroostook* and *Gannet*, as well as planes based aboard battleships and cruisers. The Blue force was considerably smaller, having only 15 cruiser-based planes and two other aircraft based on the *Wyoming*. Planes aboard the *Wyoming* were useless, however, for the battleship was not equipped with a catapult. Grimes records:

"**The Black War Diary shows that the greatest part of the air activity during Fleet Problem V involved the *Langley*. Scouting flights were conducted each day as the Black Fleet proceeded towards Guadalupe. The largest number of planes used at any one time was ten. The duration of these flights ranged from 30 minutes to two hours.**"

"**On the last day before the arrival at Guadalupe, the *Langley* received a 'well done' for the feat of launching ten planes in 13 minutes! None of these flights resulted in contacts.**"

"**On March 10, the *Langley* was ordered to have her planes ready for an 0530 takeoff the next morning. These planes were to make an aerial reconnaissance flight over the anchorage before the Black Fleet entered. This operation never took place, the problem being terminated at 0508 March 11 by the Chief Observer.**"

An *Aeromarine* lands aboard *USS Langley* (CV 1) while the ship is at anchor in 1923. *Langley*, converted from the collier *USS Jupiter* (AC 3), was placed in commission Mar. 20, 1922, at Norfolk, Va., as the Navy's first aircraft carrier. Photo courtesy U.S. Naval Institute Photo Collection. [NH93176]

Introduction of the *Langley* to Fleet operations was considered a valuable experience. As a result of this problem, the Commander-in-Chief, U.S. Fleet, recommended that the *Saratoga* and *Lexington* be completed as quickly as possible. He also urged that steps be taken to insure the development of planes

Chapter 4

The Evolution of Aircraft Carriers

of greater durability, dependability and radius, and that catapult and recovery gear aboard cruisers and battleships be further improved.

Details concerning Fleet Problem VI, conducted in 1926, are unavailable. Pertinent documents on orders, instructions and operation reports are lost. It is known, however, through the Annual Report of the Secretary of the Navy, 1926, which a combined U.S. Fleet participated in a joint Army Navy minor problem and conducted "strategically tactical exercises in the vicinity of the Canal Zone until the middle of March 1926. Fleet Problem VI was conducted during this period."

Just before Fleet Problem VII got underway in 1927, a joint Army-Navy exercise was conducted, again testing defenses of the Panama Canal. USS *Langley* provided defense against attacks on ships by land-based Army planes and was also used for spotting submarines. This exercise marks the first time an aircraft carrier was used to protect ships of the line. Battleship-based planes were used for spotting during bombardment of the Canal installations.

Canal defenses were again found weak, but again, "constructive;" planes were used in the attacks. In each of the two attacks on Miraflores Locks, only one plane was launched; it represented the attacking forces. This was not considered an effective test. Grimes noted: "In later problems when carriers were available from which attacks in force could be launched and greater reality could be introduced into maneuvers, the vital necessity for air defense of the Canal was to become even more apparent."

Typical of the early air-cooled engine carrier fighter, scout, and dive bomber types which flew from the Lexington and Saratoga in Fleet Problem IX were the Boeing F2B-1's of Bomber Squadron 2.
Source: www.history.navy.mil

Chapter 4

The Evolution of Aircraft Carriers

The seventh Fleet Problem provided more experience in carrier operations. Conducted in the Caribbean in March, Blue Fleet was given the task of escorting a large, slow, overseas convoy and was then to establish a base under enemy opposition. This Fleet was then to oppose the Black Naval Force from that base. Black's mission was to provide search and contact scouting, track submarines, and attack a large convoy accompanied by a strong escort. The *Langley* was assigned to the Blue force. Again, the converted collier-made-carrier was to provide protection for ships of the line.

On the last day of the game, Black conducted a surprise air attack—delivered by 25 land-based aircraft (Mole St. Nicholas)—against the Blue force. Shortly before this, *Langley* maintained a protective air patrol over the convoy, but discontinued it hours before the attack was pressed home. Caught unawares, *Langley's* planes were no help. Even though the problem had officially terminated by the time Black's aircraft reached Blue's ships, observers considered the attack successful, though the Commander-in-Chief scored the clumsy formation of the attacking planes.

One of the most revealing outcomes of this problem was the need to allow aircraft carriers greater latitude in maneuvering, as dictated by weather and the position of the enemy forces. Commander, Air Squadrons, also felt that he should have complete freedom of action in employing carrier-based aircraft in order to get maximum efficiency in air operations.

Fleet Problem VIII, conducted in the Hawaiian-Pacific area in April 1928, provided further experience in aircraft carrier operations and scouting patrols, *Langley*, *Aroostook* and *Gannet* again participated and again air operations were

USS Aroostook (CM-3) underway in Fleet Problems VIII - April 1928; Aroostook carried Amphibious Aircraft such as the RS-3 using a boom to launch and retrieve them.
Source: www.history.navy.mil

limited to scouting. Bad weather and heavy seas effectively limited air operations, but despite uncooperative weather, Commander-in-Chief, Battle Fleet, noted that a sufficient number of aircraft were launched from the *Langley* "to show that the use of planes from carriers for all contemplated operations is both practicable and feasible."

A Sikorsky Amphibian, similar to this RS-3, based on the Aroostook, represented all the Langley's squadrons in Fleet Problem IX. Its pilot "bombed" the Atlantic side of the Canal without opposition, landed, and informed the "enemy" of what they had just accomplished.
Source: www.history.navy.mil

Of all the Fleet Problems conducted before 1940, the next, Fleet Problem IX, undoubtedly received the most publicity. Conducted in 1929, it saw the introduction of the world's largest aircraft carriers, the *Saratoga* and *Lexington*. The problems entered their third phase. "The experience gained and the conclusions drawn," says historian Grimes of this problem, "had a marked influence on the development of fleet tactics and strategy in general, and on Naval Aviation in particular."

The Panama Canal was again chosen for the critical area under hypothetical attack. Previous exercises indicated a major weakness in defense of the Canal, protection from air attack, but this problem was to test the conclusions reached in the past by providing actual aircraft carriers and full strengths of aircraft.

The problem assumed that a war had existed between Blue (the U.S.) and two enemy nations, Black (in the Pacific) and Brown (in the Atlantic). In airpower, Blue was assigned the *Lexington*, 145 naval aircraft, and the cooperation of the U.S. Army in the Canal Zone and 37 planes based there. Black was given the *Saratoga* and the *Langley*. When it became evident that *Langley* would not

Chapter 4

The Evolution of Aircraft Carriers

complete overhaul in time for the games, the tender *Aroostook* was substituted, the single amphibian aboard representing *Langley's* 18 fighters and six scouts, though these aircraft were actually transferred to the Sara. The Brown force proved to be a paper power; neither ships, planes, nor personnel were assigned; other than in initial planning and estimates of the situation by Blue and Black, Brown ceased to be a factor in the game.

A detachment from the Blue force, including the *Lexington*, transited to the Pacific side before Black force could launch a surprise attack. On the same day, the remainder of the Blue force was to have left Hampton Roads for the Canal. It was Black's intent to destroy the Canal before this second detachment could complete the passage.

Blue's intelligence indicated that Black would attempt an attack on the Pacific side. Actually, Black planned a surprising two-pronged attack. The "squadron" aboard the *Aroostook* was to make a long-range flight, far beyond capability of return. Its attack was to be made on the Atlantic side, at the conclusion of which, the "planes" were to land and surrender. Simultaneously, *Saratoga*, accompanied by *Omaha*, was to attempt a daring tactic: take a wide, two-day swing to the south and then launch carrier-based planes for the Pacific attack. This latter demonstration was to make a profound impression on naval tacticians.

USS Rhode Island's deck with 12 inch shells for her big guns; these projectiles are more than capable of sinking a small aircraft carrier such as the USS Lexington.
Source: www.history.navy.mil

On the morning of January 25, 1929, two days before the problem was to end, the main Blue force, including the *Lexington*, came upon Black's Striking force. Black's Battleship Division Five was steaming downwind while the carrier was steaming up, preparatory to launching her planes for an air attack. The battleships opened fire and, because of the close range, would surely have sunk the *Lexington* in actual battle. For this carrier, it was a disastrous ending to her first important activity in the problem.

Umpires ruled the carrier "damaged," however, for the loss of the carrier at this early stage of the game would have had a profound restriction on

Chapter 4

The Evolution of Aircraft Carriers

Blue's capability during the coming "interesting" part of the problem. *Lexington* was instead penalized in speed; she was permitted only 18 knots.

The carrier had already launched some planes. After the attack by the battleships, the carrier, running into rain and reduced visibility, was forced to recover these aircraft under very adverse conditions. The Commander-in-Chief, U.S. Fleet Noted afterward with satisfaction: "Flight deck personnel and flying personnel alike are deserving of great credit for the manner in which squadrons came aboard on this occasion."

The *Saratoga*, in the meantime, was steaming south. She was detected by an enemy destroyer upon which she opened her eight-inch guns. This had unfortunate results. The destroyer was "sunk," but in the process, one of *Sara's* planes, a T3M, was destroyed. Spotted in the hangar deck just aft of the forward elevator and 68 feet from the muzzle of the gun, the plane suffered 36 crushed ribs and some torn fabric, directly attributable to the blast from the heavy gun. The eight-inchers were destined to be removed from the *Saratoga*, but not before WW II.

USS Lexington (CV-2) with Curtiss F6C fighters (lower right) and Martin T3M torpedo planes aft of the elevator shaft, as she arrives off San Diego, California, on her maiden cruise, 4 April 1928. The USS Lexington played an important role in this Fleet Games and in the United States becoming the world's premier operator of modern carrier's. Source: www.history.navy.mil

Later that day, the carrier encountered another Blue ship, the *Detroit*, which continued to track her through the night, supplying the Blue commander with

Chapter 4

The Evolution of Aircraft Carriers

vital information. The *Lexington* was ordered to give chase, but because of her reduced speed could not close during the night. At 0525 the next day, the Chief Observer canceled this penalty.

The 26th was an active day for the *Saratoga*, and not an altogether lucky one. Near five that morning; about 145 miles from Panama, she launched an initial attack of 70 planes against the Canal. Her aircraft in the air, the good times were over for the *Saratoga*. Because of navigational discrepancies, the carrier and the *Omaha* contacted Blue's Battleship Division 2 instead of her own Battleship Division 5. The carrier was under heavy fire at short range from three of the enemy battleships and was scored a "sinking." Had she escaped this disaster, four torpedoes "fired" from an enemy submarine at 1200 yards would have hurt her heavily and possibly sunk her. At 0656, *Sara* launched two more squadrons, at the completion of which, she had 83 aircraft in the air.

USS Saratoga steams into the wind as a Vought O2U-2 Corsair prepares to land. This photograph was taken in 1929, so it is possible that this picture was taken during Fleet Problem IX. Note Marine Corps insignia painted under the aft cockpit of the closest aircraft. Source: www.history.navy.mil
Photograph Courtesy of PHC John L. Highfill, USN (Retired).

At 0600 the same day, the *Aroostook* stopped near Jicaron Island and 17 minutes later her single plane—a Sikorsky amphibian, theoretically representing *Langley's* squadrons—took off to bomb Gatun Locks, Gatun Spillway and Coco Solo. The commander of the Blue force, though, had committed a blunder by not advising either Coco Solo or the Army of the substitution. The Sikorsky went about its mission of bombing the Atlantic side of the Canal, encountering no opposition. When he completed the job, he landed at the Atlantic terminal of the Canal, surrendered to a stunned "enemy" and confessed his accomplishment.

The Evolution of Aircraft Carriers

Saratoga's problems were not yet over. Shortly after the Chief Observer lifted the speed limitation from the *Lexington*, she launched her aircraft against her sister ship. The *Saratoga* based planes, returning from a successful raid on Miraflores and Pedro Miguel locks and the airfields at Fort Clayton and Albright were in the process of landing when the *Lexington* launched planes arrived on the scene. *Sara* "went down" for the third time.

Men in the *Lexington* had little to cheer about, though, on the morning of January 27. Friendly planes of VT-9 took off from their moorings at Balboa and stood out to attack the Black carrier and bombarding forces. About 0700 they reached a carrier and started bombing. Later, they learned it was their own ship, the *Lexington*.

"This error was understandable," said Commander-in-Chief, U.S. Fleet, "in view of the fact that *Saratoga* and *Lexington* were operating within 12 miles of each other at that time and it was not possible to distinguish markings, owing to the presence of a large number of men on turret tops. For purposes of identification, each turret top of the *Saratoga* bore two painted white stripes parallel to the axis of the guns. The turret tops of the *Lexington* were painted conventional war color." In later years, *Saratoga* was made more easily recognizable by the painting of a large stripe down the center of her stack.

USS Lexington (CV-2) Steams through an aircraft-deployed smoke screen, 26 February 1929, shortly after that year's Fleet Problem IX exercises - Source: www.history.navy.mil

Fleet Problem IX marked an outstanding achievement in Naval Aviation. It marked the first appearance of modern large carriers with the Fleet in a fleet problem. But the most significant event of this problem, and possibly in any before WW II, was the employment of *Saratoga* as a separate striking force. Its

Chapter 4

The Evolution of Aircraft Carriers

effect on the future use of carriers was immediate. In the 1930 maneuvers, a tactical unit, built around the aircraft carrier, appeared in force organization for the first time.

For many historians of naval warfare, Fleet Problem IX marked the introduction of the fast carrier task force. Regardless of its genesis, this tactical weapon was tested and refined during the war games of the Thirties. Addition of the carriers *Ranger*, *Lexington*, and *Saratoga* was to provide more flexibility and realism in future games. A discussion of them, as well as the results of the fleet problems, will be presented in the following chapter of The Evolution of Aircraft Carriers.

USS Lexington (CV-2) as she appeared in 1929, a year after she was commissioned; too late to participate in the war games of 1928, she entered the next games enthusiastically. Pilots aboard learned much from the experiences of their colleagues in the Langley. In this photo she's launching torpedo planes during Fleet Problem IX. It is amazing to consider the difficulty that these early naval aviators had in landing their aircraft on the straight decks with the island large and looming on their starboard side during final approach. As you can see, the aircraft themselves where almost as wide as the flight deck, so if they missed the arresting wires it could be a very touchy situation while full power is applied. Source: www.history.navy.mil

Chapter 5

The Evolution of Aircraft Carriers

USS Ranger (CV-4) shows forward palisade placed at the very bow of the flight decks on the early carriers with cross-deck tie-downs to protect planes and men from wind and spray. She entered the Fleet Problems in 1935 and for a small carrier with comparatively light tonnage for her day proved herself to be quite a capable weapon of war for the United States. Source: www.history.navy.mil

Chapter 5 - Last of the Fleet Problems

'The culmination of the year's operations arrives when the carriers with their squadrons participate in the annual cruise of the Fleets. On these cruises, the year's efforts to perfect the detail of aircraft operations are given the test of simulated major campaigns against possible enemies. Our efforts in the past have been crowned with a certain amount of success, but every success has only indicated new possibilities of the employment of aircraft in fleet operations and has emphasized the vital importance of continuously operating with the Fleet the maximum number of aircraft that can be carried on our surface vessels.'—RAdm. J.M. Reeves, USN, Commander, Aircraft Squadrons, Battle Fleet, 1929

Chapter 5

The Evolution of Aircraft Carriers

R Adm. Reeves described the yearlong training schedule of Naval Aviators as the Twenties came to an end:

"Concurrently with gunnery exercises, the squadrons are embarked on the aircraft carriers and they participate in the monthly exercises with the Fleet. These fleet exercises are arranged to present new and increasingly difficult problems to all arms of the Fleet and to insure the effective coordination of these arms in major fleet operations and engagements."

Vought Corsairs attached to Ranger's utility unit were typically used for scouting and observation duties during these later war games. Source: www.history.navy.mil

"It is not sufficient for one officer, Commander, Aircraft Squadrons, to be proficient in effectively employing aircraft. This knowledge must be possessed by all flag officers. To this end, aircraft on the various carriers, and the carriers themselves, are assigned from time to time in fleet exercises to the various subdivisions of the Fleet. In part of a problem, the aircraft will cooperate with destroyers; in another part, they operate offensively against destroyers; in another part, they operate with and against submarines; they operate continually with battleships and these battleship planes must continue their activities during the attack of 'hostile' aircraft. This employment of aircraft on widely differing missions reacts not only to the vast improvement of the air arm, but also and equally important, it acquaints the officers of command rank with the possibilities and effective means of employing aircraft to further the main mission of the Fleet, the destruction of the enemy."

Chapter 5

The Evolution of Aircraft Carriers

Fleet Problem Nine, conducted in 1929, created a profound impression on the tacticians of the day. In March and again in April of 1930 two more problems were presented the Fleet, both conducted in the Caribbean, and both concerned with the versatility of aircraft carriers as naval weapons. They were Fleet Problems X and XI. Fleet Problem X investigated the maneuvers necessary to gain a tactical superiority over a force of approximately the same strength and in the use of light forces and aircraft in search operations. Carriers were here defined as a complete tactical unit, operating with cruisers and destroyers as a high-speed striking force.

The Blue force, representing the U. S., was assigned both *Saratoga* and *Langley*, while the Black force, a coalition of enemy nations, operated the *Lexington*. Earliest control of the Caribbean was crucial to solving the problem.

At the outset, neither force knew exactly where his opponent was, though Black, through intelligence reports, had enough information available to assume the Blue ships would transit the Panama Canal to the Atlantic side. The ships already had.

Blue's commander considered the water too rough for the safe operation of seaplanes on the first day of the problem and was reluctant to send his carrier-based planes, for he expected to contact the Black carrier force before dark. The Black ships were in a position just north of the island of Haiti. By dawn next morning, they had moved to the west side of the island.

On the second day of the problem, the Blue commander again called off air operations because of bad weather and rough seas. Black, on the other hand, conducted extensive scouting operations while advancing to the west. Haitian-based planes scouted from daylight to dark, while *Lexington* based fighters and scouts launched every three hours for a 12-hour period.

USS Saratoga and USS Lexington off Virginia Capes along with 11 battleships just after completing Fleet Problem X. A crewmember of the mammoth dirigible USS Los Angeles (ZR-3) took this photograph. Source: Hampton Roads Naval Museum

Chapter 5
The Evolution of Aircraft Carriers

Weather improved on the third day and the Blue commander ordered his carrier planes launched. Still neither side had any idea where the opponent was. This status continued through the fourth day, and it was not until the fifth that contact finally was made.

Saratoga was spotted by *Lexington* aircraft and as a result of the attack that followed, *Sara*'s flight deck was damaged. Before her planes could be repositioned for launching off the usable end of her deck, Saratoga suffered another and finishing attack. *Lexington* next turned her attention to the *Langley* and in two flights of first 15 and then 12 planes successfully placed the converted collier's flight deck out of commission.

Next, USS *Litchfield,* one of *Saratoga's* plane guards, was dive-bombed and placed out of action. Blue's battleships then felt the effects of *Lexington's* planes with the result that the *West Virginia* suffered the destruction of two anti-aircraft guns, the *California* lost an observation plane on deck, injury or death to personnel, foretop material damaged, and a 15 per cent reduction in main battery fire; and the *New Mexico,* lost four AA guns as well as an observation plane still on one of the ship's turrets. Neither *Saratoga* nor *Langley* took part in the main action that followed the destruction of their flight decks.

The USS Langley's role in the war games became decreasingly important as new aircraft carriers were added to the Fleet. Aircraft such as this Curtiss F6C became standard on carriers.
Source: wp.scn.ru/en

At its conclusion, Fleet Problem X demonstrated the suddenness with which on engagement could be completely reversed by the use of air power. Scouting planes and scouting operations were also scored, the planes found wanting in range and the scout pilots unable to bomb carrier decks when contact was made.

A month later, Fleet Problem XI investigated further the limitations of scouting planes as well as their most effective use. After the game, it was recommended that scouting squadrons should be increased to 18 planes and that a more suitable scouting plane be developed. It was felt that better flotation was needed for amphibians and that a greatly increased range for carrier-based scouts, as well as the ability to take off with a short run was necessary. Among desirable secondary characteristics were small size, folding wings, and high speed, even at the cost of ceiling and armament. It was also recommended that semi-permanent task groups be organized, each consisting of one large aircraft carrier, a division of cruisers, and a division of destroyers. These ships were to be trained as a unit in frequent exercises.

The 1931 Fleet Problem (XII), conducted in the Pacific-Panama Bay area, had among its tasks exercises in strategic scouting, in the employment of carriers and light cruisers, and refueling at sea.

USS Los Angeles moored to USS Patoka during Fleet Problem XII 1931. The U.S. Navy had several dirigibles like the USS Los Angeles. It was hard to find a proper use for these due to some very poor characteristics. These included a predisposition to explode when hit with tracer fire from machine guns and their poor handling in strong winds. Source: www.history.navy.mil

Primarily, this problem dealt with actions between a fleet strong in aircraft and weak in battleships, and in a reverse situation where the fleet was weak in aircraft. At its conclusion, it was considered that two cruisers and two destroyers were minimum protection for an aircraft carrier in a carrier group. Further, the commander of that group should be stationed in the air craft carrier, rather than in a cruiser or destroyer, so that he could fully understand the mission of that group and obtain its quickest cooperation. Also, it was pointed out, escorting vessels must maintain the speed and proportionate fuel capacity of the carrier.

Chapter 5

The Evolution of Aircraft Carriers

At the end of the problem, the three carriers transited the Canal and headed for Cuban waters and more exercises. On the last day of March, Capt. Ernest J. King, commanding *Lexington*, was ordered to assist Navy and Marine units in relief operations in Nicaragua. An earthquake had destroyed most of the city of Managua. When *Lexington* launched five aircraft with medical personnel and supplies aboard, in addition to provisions, she inaugurated carrier aircraft relief operations in the U.S. Navy. This was to become a frequent peacetime mission.

Canal Zone based aircraft, such as this early Vought O2U-4 Corsair floatplane participated in the Fleet Problem games primarily as scouts. This O2U-4 is being launched from a cruiser. Prior to the development of radar, cruiser-deployed scouting aircraft were vital for the trade protection role of the cruiser force. Source: www.history.navy.mil

During Fleet Problem XIII held in the Pacific-West Coast area in 1932, the vulnerability of submarines to air detection and attack, at that time, was clearly demonstrated. Four out of five submarines of one force, assigned scouting missions, were detected by land and carrier-based planes and "sunk." C.O.'s of these submarines reported their own vulnerability when operating in an aircraft-screened area.

Aircraft carriers assigned to the problem were forced to exercise in widely separated areas of the Pacific Ocean. RAdm H. E. Yarnell, who commanded the U.S. aircraft during the exercise, noted that in event of actual war in the Pacific, the number of aircraft carriers on hand would be totally inadequate to meet the needs.

Chapter 5

The Evolution of Aircraft Carriers

Also, the admiral pointed out, this problem was not greatly dissimilar from all other problems conducted in the past, in that when one aircraft carrier was assigned to each of the forces in the war games, each of the forces invariably made the destruction of the other's carrier the prime tactic. This resulted in both forces losing their carriers early in the game.

It was therefore obvious, he repeated, that the side with the greater number of carriers has a tremendous advantage. In time of war, this would be critical. He suggested that at least six or eight more aircraft carriers be added to the Navy's inventory.

The next problem, XIV, was conducted in the same area the next year, 1933. Its conditions were that "during preparation for escorting an expeditionary force overseas in a campaign, an outlying possession was industrial, military and mobilization centers of a long coast line were threatened by carrier raids."

The Blue force was to protect the West Coast while Black was ordered to make at least one raid in the San Diego-Dan Pedro, San Francisco and Puget Sound areas. Black divided it force into three groups. Its Northern Carrier Group was to raid San Francisco and then proceed to the Puget Sound to the north. The Southern Carrier Group was to raid San Pedro and then San Francisco, rendezvousing later with Black's Support Group.

USS Saratoga preparing to launch several squadrons of Boeing F3B-1 light fighter bombers. Propellers are spinning; pilots and crews are at the ready
Source: www.history.navy.mil

The first four days were uneventful. On the firth day, a *Lexington*-based plane of the Northern Group spotted an enemy submarine, causing the carrier to change formation for the approach to the launching point of the raid. Weather worsened, forcing the suspension of flight operations. Early the next morning, as *Lexington* warmed up her planes, a Blue battleship was sighted at a 4500-yard range. As the carrier tried to escape, a second enemy battleship came into view and the Northern Carrier Group was declared out of the action, caught unexpectedly between two enemy battleships as close range.

Chapter 5

The Evolution of Aircraft Carriers

The Southern Carrier Group had better luck. On the seventh day of the problem, Saratoga-based planes successfully launched the attack. Black reported that 12 scouts had attacked the oil refinery at Venice with 24 100 – lb. bombs, five scouts attacked a power house at Long Beach with ten equally powerful bombs, encountering no enemy force and sustaining no losses. The force lost three bombers to the enemy's two fighters during an 18-bomber attack on an enemy transport, an oil field at El Segundo and docks at Long Beach. Saratoga sustained slight damage. The force moved north for the San Francisco raid.

When she arrived in the San Francisco area, *Saratoga* launched her planes. Before she completed, aircraft from the cruiser *Richmond* and the carrier *Langley* bombed her flight deck. After *Sara*'s planes returned from the raid, 37 percent of her flight deck was assessed damaged, 36 planes lost, and her flight deck out of commission for 38 hours. The CV-2 aircraft had succeeded in making a dive bombing attack on the *Langley*, temporarily disabling her flight deck, and attacked Crissy Field, San Francisco docks, San Andreas reservoir, and the dry-dock at Hunter's Point.

This exercise underscored the urgent requirement for the development of better planes, particularly carrier bomber and torpedo planes. Adm. Yarnell again pleaded for three additional 18,000-ton carriers which were permitted under existing treaties.

In the period 1933-34, the Fleet conducted a series of 20 tactical exercises. The last three of these comprised Fleet Problem XV, which also proved the last of the war games of the three-carrier period.

In his official monograph "Aviation in the Fleet Exercises, 1911-1939," historian LCdr. James M. Grimes, USNR, described the war games: "The primary effort of the Commander-in-Chief when drawing them up had been to introduce realism into fleet tactics and to simulate as nearly as possible actual wartime operations. For this reason, the opposing fleets represented actual navies of the period. Carrier operations were extensive throughout the problem."

"There were several important results of Fleet Problem XV as regards the development of Naval Aviation. The most important, perhaps, was the realization brought out by air operations during the problem, that if the carrier was to be the offensive weapon it was considered to be, carrier-based planes would have to be so armed that they could carry the offensive to the enemy."

"It was seen that planes carrying 100-lb. bombs were obsolete and of little use against an enemy force equipped with planes capable of carrying 500 and 1000 lb. bombs. The Commander-in-Chief, in his remarks at the critique held on Fleet Problem XV, stated that at least three-fourths of the carrier-based planes should be so equipped."

Chapter 5

The Evolution of Aircraft Carriers

BOEING F4B-4's were famous carrier fighters. One is now in the National Air and Space Museum in Washington DC, USA - Source: San Diego Air and Space Museum Photo Archives.

USS *Ranger* joined the Fleet for the next war game, Fleet Problem XVI, conducted in 1935. Actually, this game consisted of five separate exercises, none of them related, spread over the Pacific from the Aleutians to Midway, to Hawaii. Both the Army and Coast Guard participated.

The major air operations took place during the third phase of the problem. Unfortunately, these were marred by a series of plane and personnel casualties that seriously affected later air and sea operation. Although valuable experience was obtained in mass flight of patrol squadrons, nothing of significance developed in the operation of aircraft carriers.

Fleet Problem XVII was conducted in the Panama-Pacific area in 1936. The exercises (again five) saw extensive use of patrol planes and the effective use of automatic pilot, but there was no major contribution to, or effect on, the evolution of carriers, either in design or tactics.

The question of proper employment of aircraft carriers was brought up again in Fleet Problem XVIII of 1937: Should they operate with the main body of a fleet or should they operate at a distance?

Black's aircraft commander held that a carrier tied down to a slow main body formation was certain to be destroyed. "Once an enemy carrier is within striking distance of our Fleet," he said, "no security remains until it, its squadrons, or

Chapter 5
The Evolution of Aircraft Carriers

both are destroyed, and our carriers, if with the main body, are at a tremendous initial disadvantage in conducting necessary operations."

But his force commander took a different view. He felt that carriers should be an integral part of the main body and defended his decision to employ them in such a way, as he did in this problem. He suggested that *Ranger*, because of her small size could provide scouting and spotting with less chance of being detected. He hoped that when *Yorktown* and *Enterprise* joined the Fleet, such an employment of *Ranger* might be possible.

Fleet problem XIX was the last of the *Ranger* phase of the war games. It was conducted in 1938 and consisted of Parts II, V, and XI of the Annual Fleet Exercises.

In the first phase, the outstanding performance was a long-range San Diego-based patrol plane bomber attack which successfully eliminated *Lexington* as a carrier unit in the game.

With the entry of USS Enterprise (CV-6), above, and USS Yorktown (CV-5) into the war games in 1939, the fleet problems entered their final phase. Tactics and ships' operations were refined in the tense years immediately prior to the United States' entry in WW II. Source: navysource.org

The notable development of the second phase of the war game, Part V, was an attack on Pearl Harbor, launched from *Saratoga* some 1000 miles off the coast of Oahu. *Sara*'s recon group flew over Lahaina area, photographing beaches and reporting the enemy's strength there.

At the same time, *Saratoga* sent an attack group which bombed Fleet Air Base, Hickam Field, Wheeler Field, Wailupe Radio Station, and returned to the carrier. This tactic was to be employed by the Japanese some three years later, in December 1941.

Chapter 5

The Evolution of Aircraft Carriers

In phase three (Part XI), the outstanding air operation was an unopposed air attack by *Lexington* and *Saratoga* based planes launched against Mare Island and Alameda.

"Excellent experience was provided in planning and executing a fast carrier task force attack against shore objective," says Grimes. "The problem of defending a coast line or even an isolated portion thereof, against fast enemy raiding forces equipped with large carriers and protected by powerful surface ships was seen to be one difficult of solution."

Yorktown and *Enterprise* entered into the 1939 exercises of Fleet Problem XX, which were conducted in the Caribbean area and off the northeast coast of South America. The war games entered their final phase. Neither *Langley* nor *Saratoga* participated.

USS Enterprise (CV-6) was the sixth aircraft carrier of the United States Navy. She was a ship of the Yorktown class and was launched in 1936. The Enterprise was one of only three American carriers commissioned prior to the Second World War to survive the war as she took part in more major military engagements of the war against Japan than did any other U.S. ship. She participated in the Battle of Midway, the Battle of the Eastern Solomon's, the Battle of the Santa Cruz Islands, various other air-sea engagements during the Guadalcanal campaign, etc. She was decommissioned in 1947. Source: History Wars Weapons

The USS Enterprise (CV-6) was built by Newport News Shipbuilding. She was laid down in 1934 and launched in 1936. This aircraft carrier was 824 feet long, displaced 19,500 tons of water, and had a draft of 25 feet. The USS Enterprise (CV-6) had 9 Babcock & Wilcox boilers, and was propelled by 4 Parsons geared turbines. She had 3 elevators, two flight hydraulic catapults, and could carry 90 aircraft. On 8 April 1942, she departed to meet the USS Hornet and sail west escorting Hornet on the mission to launch 16 Army B-25 Mitchells in the "Doolittle Raid" on Tokyo. Source: History Wars Weapons

As a result of this game, reports indicated that carrier operations reached a new peak of efficiency; particular credit was given the two new carriers which, despite inexperience, contributed significantly to the success of the problem. These exercises studied employment of planes and carriers in connection with convoy escort, development of coordinating measures between aircraft and destroyers for anti-submarine defense, attack on mobile patrol plane bases,

Chapter 5

The Evolution of Aircraft Carriers

scouting and attack by patrol planes, defense of surface ships against aircraft attack, and trial of various forms of evasion tactics against attacking aircraft and submarines.

The last war game XXI was played in 1940 in the Hawaiian-Pacific area. It consisted of two separate exercises. The Historian Grimes describes them:

"The first exercise was designed to afford training in making estimates and plans; in scouting and screening; in the coordination of various types of fighting units; in employing standard and fleet dispositions; and finally to train the opposing forces in decisive engagement."

"The second major exercise of the problem was designed to afford training in scouting, screening, communications, coordination of types, protection of a convoy, seizure of advanced bases and finally, decisive engagement."

Between the two major parts of the problem were two minor exercises in which air operations played a major part: Fleet Joint Air Exercise 114A and Fleet Exercise 114. Exercise 114A underscored the need for greater cooperation between the Army and Navy in organizing the defense of the Hawaiian area. Exercise 114 compared patrol plane attacks on surface units with use of planes in high altitude tracking. The former proved the planes vulnerable, while the latter met with great success.

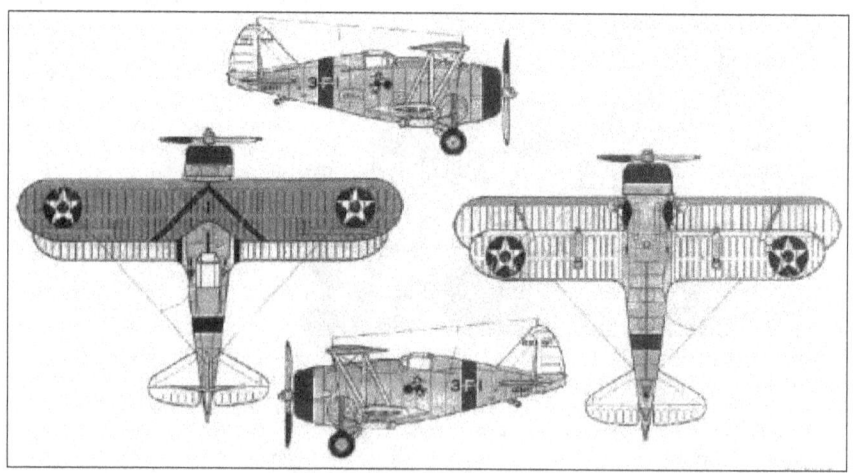

Grumman F3F biplanes were a mainstay of carrier fighters until shortly before WW II. Source: blueprints.com

Few new difficulties emerged from this war game. Reiterated was the question of latitude given carrier commanders by force commanders. *Yorktown*'s commanding officer stated his belief that success could best be achieved when aircraft personnel in carriers operated under a broad directive. The exercise proved again—as it did in Fleet Exercise 114—that low level horizontal

Chapter 5

The Evolution of Aircraft Carriers

bombing attacks had little chance of success— especially against a ship that was not otherwise engaged.

By 1940, the war games were halted. Although one was planned for the next year, worsening of world tensions caused their cessation. Various tactical exercises were held instead.

USS Yorktown (CV-5) at the Newport News Shipbuilding and Dry Dock Company, Newport News, Virginia, in June 1937, while preparing for sea trials. Source: www.history.navy.mil

The USS Yorktown was the namesake of her class. She sailed the Pacific and Atlantic Oceans and sunk on the afternoon of 4 June 1942, after her aircraft had sunk the I.J.N. aircraft carrier Soryu. She was hit by two torpedoes fired by Japanese Zeros and sank near Midway Atoll.

The Yorktown was laid down on 21 May 1934, at Newport News, Virginia, by the Newport News Shipbuilding and Drydock Co, sponsored by First Lady Eleanor Roosevelt. She was launched on 4 April 1936, and was commissioned at the Naval Operating Base, Norfolk, Virginia, on 30 September 1937. Her first commander was Capt. Ernest D. McWhorter.

The Yorktown was 770 ft. long, displaced 19,800 tons, and had a draft of 25 ft. 11.5 in (7.91 m). She was propelled by 4 Parsons geared turbines making 120,000 shp, a top speed of 32.5 knots. The Yorktown could carry 90 aircraft, and was fitted with (8) 5 inch / 38 Cal guns, (4) Quad 1.1 in / 75 Cal guns, and (30) 20 mm cannon.

Naval Aviation grew with the war games. The first phase—the pre-aircraft carrier years-employed "constructive" carriers and merely indicated to the Navy the potentials of this new weapon. The *Langley* phase was an informative one, but this was more an experimental ship than an aircraft carrier. The games

Chapter 5

The Evolution of Aircraft Carriers

reached fruition with the addition of the *Lexington* and *Saratoga* in Fleet Problem IX. It saw the employment of an aircraft carrier as a separate striking force and introduced a new tactic in the book of naval strategy. The *Ranger* phase showed the potentials of small aircraft carriers, employed with telling effect in WW II. And the final phase, the addition of *Yorktown* and *Enterprise*, increased and refined carrier operations in the critical years prior to WW II.

Chapter 6

The Evolution of Aircraft Carriers

*Imperial Japanese Navy AC Carrier Junvo at Sasebo Japan Sep 1945.
Source: www.history.navy.mil*

Chapter 6 - The Japanese Developments

'In the last analysis, the success or failure of our entire strategy in the Pacific will be determined by whether or not we succeed in destroying the U.S. Fleet, more particularly, its carrier task forces.'—Adm. Isoroku Yamamoto, IJN, 1942. 'I think our principal teacher in respect to the necessity of emphasizing aircraft carriers was the American Navy. We had no teachers to speak of besides the United States in respect to the aircraft themselves and to the method of their employment. We were doing our utmost all the time to catch up with the United States.' — FAdm. Osami Nagano, Imperial Japanese Navy (IJN), 1945.

By Christmas Eve 1921, the Washington Disarmament Conference had already been going on for a month and a half. Participating were Great Britain, Japan, France, Italy, and the United States. It was on this day that Great Britain refused any limitation on auxiliary vessels, in view of France's demand for 90,000 tons in submarines. The delegates then began to consider confining the treaty to capital ships and aircraft carriers.

The Washington Naval Treaty, signed February 6, 1922, established a tonnage ratio of 5-5-3 for the capital ships of Great Britain, the United States, and Japan, respectively, assigning a smaller tonnage to France and Italy. The same ratio for aircraft carriers was set, with an overall limitation of 135,000 tons each for Great Britain and the U. S., and 81,000 tons for Japan. It also limited any new carrier to 27,000 tons, with a provision that, if total carrier tonnage were not

Chapter 6
The Evolution of Aircraft Carriers

thereby exceeded, nations could build two carriers of not more than 33,000 tons each, or obtain them by converting existing or partially constructed ships which would otherwise be scrapped by the treaty.

December 27 that year, Japan commissioned its first aircraft carrier, the *Hosho* ("Flying Phoenix"). This was a remarkable hoku bokan (literally, mother ship for aircraft). Though the British were the first to operate aircraft onto and off a ship especially designed for that use, their first aircraft carriers were conversions. *Hosho* was a carrier from the keel, the first of its kind completed in any navy of the world.

Laid down in 1919 at the Asano Shipbuilding Co. of Tsurumi, the ship was fitted out at Yokosuka Navy Yard at a standard displacement of 7470 tons, a speed of 25 knots, with the capability of handling six bombers (plus four reserve), five fighters (in addition to two in reserve), and four reconnaissance planes, a total of 21 aircraft.

Japanese Aircraft Carrier Hosho, 1922-1947; seen here off Kure, Japan, 13 October 1945 Hosho was typical of the early Japanese aircraft carriers; she was thin, had a flat top and was quite fast. Source: www.history.navy.mil

Hosho was indeed a strange looking craft. She was all flying deck. Originally, she had an island structure and a tripod mast, but either because of the small width of her flying deck (and its attending hazards) or because some turbulence might have been caused by it, the island was taken off.

The carrier sported three funnels on the starboard side. These were of the hinged type, held upright when not in use, and swung outboard to provide additional safety from stack gas. Later, they were placed in a fixed position, bending aft and slightly downward.

Chapter 6

The Evolution of Aircraft Carriers

Under The Washington Naval Treaty, Japan converted a battle cruiser to aircraft carrier characteristics. In 1928, that country's 2nd carrier was completed and named Akagi, after a mountain. Photograph was taken by a Japanese pilot or crew at sea during the summer of 1941, with three Mitsubishi A6M "Zero" fighters parked forward and her island positioned on her port, amidships. Photograph donated by Kazutoshi Hando, 1970. Source: www.history.navy.mil

Hosho's original armament consisted of from 14cm single mount guns and two 8cm single mount high angle guns. At the outbreak of WW II, her high angle guns were replaced by four 25mm twin mount machine guns. Later, the 14cm guns were removed and 25mm double or single mount machine guns were added.

Before continuing with Japanese development, an explanation of the naming of their aircraft carriers is in order.

"**Transliteration of the names of Japanese aircraft carriers into American equivalents is a pretty risky business,**" said Mr. Roger Pineau, a frequently published writer on the Japanese Navy after World War II. "**It becomes misleading. The names should be treated as such and should not be taken too literally. For instance, when we speak of astronaut Carpenter, we don't visualize a man walking around with hammer and saw in hand.**"

Mr. Chris Beilstein, another expert on Japanese aircraft carriers, concurs. "**The *Shokaku* becomes 'Flying Crane,' for that is the closest we can translate the original Japanese. The first Japanese CV's carried names of mountains and provinces. These, in turn, were frequently named after mythological characters. *Shokaku*, for example, could have been a flying crane in an age-old story, a crane that was named Shokaku. This is very much like our real life Misty, the wild horse. Certainly, to translate 'Misty' to literal Japanese would be meaningless to them, or at best, misleading. It would be more accurate to translate it 'Wild Horse.' Thus, 'Misty,' to the**

Chapter 6

The Evolution of Aircraft Carriers

Japanese, would mean 'Wild Horse,' just as we would erroneously translate *Shokaku* as 'Flying Crane.'

"Think of the problem in transliterating Shangri La into Japanese," said Mr. Pineau. "To paint the picture accurately, it would be necessary to describe Hilton's book and then go into President Roosevelt's fascination with it. That would be rather difficult to do in one or two words. Perhaps the closest would be 'Paradise of the Ageless'—and this would, in the Japanese mind, seem a pretty silly thing to name an aircraft carrier.

"But transliteration has a very real value—especially to those who have difficulty in pronouncing Japanese words. Many competent researchers don't even speak the language. The transliteration is a handy reference point, but should not be taken seriously, at face value."

Japanese Naval Aviation dates back to 1912 when the Navy sent officer trainees to the U.S., Great Britain, and France. They returned from France with two Farman seaplanes, and from the U.S. with two Curtiss seaplanes. A beach on the west side of Tokyo Bay, Oppama, was selected as a site for a seadrome in the fall of that year and placed into commission. The first class at Oppama consisted of four officers and 100 men.

From 1912 to 1917, ¥3-400,000 (about $150-200,000) was allotted to the fledgling air arm. In 1918, this sum was increased to ¥1 million (about $500,000), and the next year to ¥2 million.

The first landing on the *Hosho* was made by a British civilian, a Mr. Jourdan, on February 22, 1923. States the Japanese Year Book of 1924-25: "our naval flight officers are making similar experiments with good results."

(In chronological comparison, Eugene Ely landed on a platform on the armored cruiser USS *Pennsylvania* January 18, 1911; USS *Langley*, the U.S. Navy's first aircraft carrier, a converted collier, was commissioned March 20, 1922; the first U.S. aircraft carrier built as such, from the keel, USS *Ranger*, was not commissioned until June 4, 1934.)

Code Named "Claude," Mitsubishi A5M4 Type 96 fighters replaced Japanese Navy's Type 90's.
Source: forummarine.forumactif.com

Chapter 6

The Evolution of Aircraft Carriers

A naval expansion program, decided upon in 1920, was completed by March 1923. Under the limitations set by the Washington Naval Treaty, Japan turned her attention to the conversion of the battle cruiser (then eight months under construction at the Kure Naval Arsenal). This, in 1928, became Japan's second aircraft carrier, the *Akagi* ("Red Castle," actually the name of a Japanese mountain). *Akagi* displaced over 30,000 tons standard when completed, had a speed of 31 knots, and carried 60 aircraft. She was armed with (10) eight-inch and (12) 4.7-inch guns.

A sister ship, the *Amagi* ("Heavenly Castle"), was also scheduled for conversion at that time, but sustained severe damage in the earthquake of September 1, 1923. She was scrapped in July 1924 at Yokosuka. In her place, Japan converted the *Kaga* (the name of an old Japanese province) to an aircraft carrier. Originally, she was laid down as a 39,000-ton battleship, but was scheduled for the scrap pile as a result of agreed disarmament limitations. Conversion was completed in 1928 and she was commissioned the following year. The 1929 Japanese Year Book states of *Akagi* and *Kaga*:

IJN Aircraft Carrier Akagi pictured fully laden with aircraft shortly after her reconstruction to add a small starboard-side island.
Source: hazegray.org

"They are the pride of the Japanese Navy, and though slightly inferior to the *Saratoga* of the U.S. Navy in respect of speed, the *Akagi* surpasses the other in point of the range of her high angle guns, of which she carries 12 4.7-inchers. The *Hosho* [is] by far smaller than the *Akagi*, but in the mode of construction [it possesses] special features of [its] own. The completion of the *Kaga*, only second to the *Akagi*, is a powerful addition to the Japanese Navy."

Kaga was reported as displacing 26,900 tons standard, but actually displaced over 30,000 tons, had a speed of 27 knots and carried 60 aircraft.

As the signatories of the Washington Naval Treaty reconvened in London in 1930, Japanese naval officers began to chafe under the ship construction restrictions imposed upon their nation. At that time, the armed forces were unpopular with the liberal government in power. Final decision on the size of the Navy lay in the competence of the civilian government. Most career officers were hostile to the treaty; those officers, who supported the civilian government in the bitter fight that ensued concerning ratification of the 1930 London Treaty,

Chapter 6

The Evolution of Aircraft Carriers

were either forced to resign within the next few years or were placed in unimportant posts. Militarists, ascending in power, referred contemptuously to the ratification as "the May 15th Affair."

The London Treaty carried forward the general limitations of the earlier Washington agreement and provided for further reductions of naval armament. Under terms applicable to Naval Aviation, the definition of an aircraft carrier was broadened to include ships of any tonnage designed primarily for aircraft operations. It was agreed that installation of a landing-on or flying-off platform on a warship designed and used primarily for other purposes would not make that ship an aircraft carrier. It also stipulated that no capital ship in existence on April 1, 1930 would be fitted with such a platform or deck.

IJN Aircraft Carrier Ryujo during sea trials in 1933. She possessed no island so she had a large deck for her size. Source: www.history.mil

The Japanese Navy expanded rapidly after 1930, at such a rate that it became necessary to conscript men. In 1931, a replenishment plan was authorized to the Navy, permitting it to complete construction of the *Ryujo* ("Galloping Dragon"), a small aircraft carrier of about 10,000 tons laid down in 1929. It was completed in 1933, its limited deck free of an obstructive island. *Ryujo* had a speed of 29 knots, carried 36 aircraft, and was armed with 12 five-inch guns. She was Japan's fourth aircraft carrier. In June 1934, USS *Ranger* became the United States Navy's fourth carrier.

In 1932, naval authorities referred a second naval replenishment plan to the Ministry of Finance for study. The plan called for a total expenditure of ¥460,000,000 (about $230 million), covering the construction of one aircraft carrier of 8000 tons, other capital and auxiliary ships, and the establishment of eight flying corps on land: all this to be completed by the end of 1936. This aircraft carrier was never built.

In 1934, preliminary disarmament conferences were held in London. Congress had already passed and President Roosevelt authorized an act that popularly became known as the Vinson Trammell Act. This permitted the U.S. to construct naval ships to the tonnage limitations prescribed by the previous Washington and London Naval Treaties. Under this authorization, USS *Wasp* (CV-7) was laid down in 1936.

Chapter 6

The Evolution of Aircraft Carriers

Japanese militarists were not eager to continue in the disarmament pacts; Wrote U.S. Ambassador to Japan, Joseph C. Grew, "Japanese attitude toward the coming Naval Conference in 1935 London Treaty is intensely unpopular among the Japanese Naval officers high and low;" and in separate correspondence, "The situation is entirely different from that in 1930.

IJN Aircraft Carrier Ryujo was so top heavy that she was incredibly unstable in heavy weather. She was brought back into the Yokohama shipyard to increase the ballast and blistered to increase her beam improve her stability and seaworthiness. Source: Maritime History Museum, Kure, Japan

Under present conditions the Navy alone will have the final say as to the size of the Imperial Japanese Navy." It boiled down to this: Japan wanted quantitative as well as qualitative parity in ship power, equal to the United States and Great Britain. The 5-5-3 ratio was no longer acceptable. Neither the U.S. nor Britain favored such an increase in Japanese strength, for, granted equality in armored ships, Japan would be the major power in the Pacific, greater than the U.S. and Great Britain combined; their Fleets were divided geographically. Japan

Chapter 6

The Evolution of Aircraft Carriers

persisted. The Japanese Year Book of 1935 enumerated that country's "official" reasoning:

"The progress and development made recently in battleships, aeroplanes, etc., have made it extremely difficult to effectuate defence operations."

"(2) The remarkable increases in the air forces of the U.S.S.R. and China, and the revival of the Far Eastern naval forces of the former."

"(3) The establishment of the naval port of Singapore by Great Britain, and the extension and strengthening of the naval port of Hawaii by the U.S.A. have had a great effect on the naval plan of operations in Far Eastern waters."

"(4) The birth of Manchukuo [independence of Manchuria, February 18, 1932] has brought for the vast changes in Far Eastern policies. It has increased the responsibility of the Japanese Empire as the stabilizing power in the Far East."

These were political arguments the world's two top naval powers could not buy. But Japan was adamant, refused compromise and, on December 29, 1934, gave the required two years' formal notice that after December 31, 1936, she would no longer be bound by the terms of the Washington and London Naval Treaties. Her act of abrogation unleashed the restraints on international shipbuilding.

Two more aircraft carriers were laid down in Japanese ways in 1934 and 1936, the *Soryu* ("Blue Dragon") and *Hiryu* ("Flying Dragon"). *Soryu* displaced about 18,000 tons standard, had a speed of 34.5 knots, and handled 63 aircraft. *Hiryu* was heavier, 18,500 tons standard, and had a speed of 34.3 knots. Officially, both ships were carried on the books as of 10,050 tons standard; the true tonnage was not revealed until after WW II. Both ships carried the same number of planes and had the same armament, 12 five-inch guns.

The Soryu Class was first laid down in 1934 and 1936, displacing about 18,000 tons standard, at a speed of 34 knots. The Soryu had her island on the starboard (conventional) side. She, with other IJN aircraft carriers, participated in the Dec. 7, 1941 Pearl Harbor raid. Source: www.history.navy.mil

Chapter 6

The Evolution of Aircraft Carriers

Japanese Zero, Mitsubishi Type O were an important part of Japan's aircraft carrier fleet in the years before and during WW II. They were proficient fighter aircraft capable of very tight turns that could give some U.S. pilots a tough time in a dog fight if the IJN pilot had serious skills, and many of them did. This painting is of unknown origin. Source: www.history.navy.mil

It was sometime between 1935 and 1937 that naval ship designers configured carriers to provide a surprising technical innovation. *Akagi* and *Kaga* underwent major modernization at this time. The lower flight decks were suppressed, the upper flight decks were extended forward, and the eight-inch gun turrets and mountings were reduced in *Akagi* from ten to six, while *Kaga* replaced her 12 4.7-inch guns with 16 five-inchers. *Kaga*'s unwieldy funnels were also reduced. The modernization of *Kaga*, which included new machinery, added about 1½ knots to her speed, giving her 28.3, but *Akagi*'s modernization cost her several knots, bringing her down to 28.

But the startling innovation was the introduction of small islands on the port side of the carriers *Akagi* and *Hiryu*. The remaining carriers had islands on the starboard (standard) side—of those that had them at all. Strategists planned to use these carriers in a formation that was unique. The lead carriers in the basic formation were to be the port-islanded *Hiryu* and *Akagi*, followed by the *Soryu* and *Kaga*. This would supposedly allow for a more compact formation with non-conflicting aircraft traffic patterns. This formation was used in the Battle of Midway.

Chapter 6

The Evolution of Aircraft Carriers

Japan's next venture into aircraft carrier construction was the *Shokaku* ("Flying Crane") and *Zuikaku* ("Lucky Crane"). These carriers were kept fairly well under wraps, insofar as specifications are concerned. They were authorized under the very ambitious Fleet Replenishment Program of 1937, the same program under which the famed super battleships Yamato and Musashi were built.

IJN Aircraft Carrier Shokaku readies her dozens of Mitsubishi "Zeros" to launch. Picture was taken by a Japanese seaman while on board in 1941 shortly after her sea trials. Source: www.history.mil

Shokaku was laid down December 12, 1937 at the Yokosuka Navy Yard, while *Zuikaku* was started at Kawasaki Dockyard May 25, 1938. Basically, the ships had similar specifications. They displaced 25,675 tons standard, had a designed speed of 34.2 knots, carried 16 five-inch guns in twin mounts, and could carry up to 84 aircraft, although a normal complement was 73. There were no major differences between the ships. *Zuikaku*, however, was fitted with a bulbous bow, the first Japanese warship so designed. *Shokuku* was launched June 1, 1939, and completed August 8, 1941; *Zuikaku* was launched November 27, 1939, and completed September 25, 1941.

Completion of both carriers was delayed when the original funnel arrangement was changed in mid-construction by the Aeronautical Headquarters. As designed, the funnels were to appear one on each side of the island bridge, fore and aft on the starboard side. This was changed by placing the two funnels immediately aft of the island.

Chapter 6

The Evolution of Aircraft Carriers

The Japanese did not give either ship much publicity. Both ships, *Zuikaku* and *Shokaku*, were to figure prominently in most sea battles of WW II involving naval air. Their design was based on the best material gathered from experiences in *Akagi*, *Kaga*, and the *Soryu* types. Later Japanese carriers (i.e., multiple ship design classes) were constructed in two groups: the large to be like *Taiho* (with armored flight deck), and the medium to be repeats of the *Soryu* class. *Zuikaku* and *Shokaku* comprised an entire class.

The Shokaku Class consisted of two carriers, Shokaku (above) and Zuikaku. They were authorized under the Fleet Replenishment Program of 1937, displacing 25,675 tons standard. Zuikaku was the first with a bulbous bow configuration. Both were completed in 1941. Source: www.history.navy.mil

Japan's next aircraft carrier was a conversion. In 1936 the submarine depot ship *Takasaki* was under construction. While she was still in the ways, the decision was made to complete the ship as a carrier. Work on this project was not started until January 1940, but was completed in December that year. The carrier was renamed *Zuiho* ("Happy Phoenix"). She displaced 11,200 tons standard, sailed at 28 knots, and carried 30 aircraft. She was armed with eight five inch guns.

A sister ship, *Shoho* ("Lucky Phoenix"), converted between January 1941 and January 1942, was originally named *Tsurugisaki*, launched as a submarine depot ship in 1934. *Zuiho* and *Shoho* particulars were similar.

Other aircraft carriers were under construction or conversion. At least 15 more would be commissioned during the war years, produced in growing restrictions of limited materials, and, after the Battle of Midway in 1942, in desperation.

Chapter 6

The Evolution of Aircraft Carriers

In The Five-Year period preceding December 7, 1941, the military of Japan grew stronger in power. March 1936 the cabinet was dominated by men in uniform and the development of heavy industry was pushed. An extraordinarily ambitious and successful expansion of the Navy was launched in 1937, the same year hostilities broke between Japan and China. That same year, the *Panay* was sunk. In 1938, the National Mobilization Bill was passed. In September 1940, Germany, Italy and Japan concluded a three-power pact. November 1941, Japanese Prime minister, Gen. Hideki Tojo, stated that British and American influence must be eliminated from the Orient.

Imperial Japanese Navy Admiral Isoroku Yamamoto at his charts in 1941. Yamamoto was the mastermind behind the attack on Pearl Harbor and was a born warrior and leader of men
Source: www.history.navy.mil

The Japanese Navy had been conducting intensive training of its officers and men during this period. Most of the training, including war games, was conducted in out-of-the-way gulfs and in the stormy northern reaches of the Pacific. The men were hardened by the elements and drilled continuously. To

Chapter 6

The Evolution of Aircraft Carriers

avoid antagonizing the Japanese, the U.S. Navy at the same time was instructed to hold all of its fleet problems in the less satisfactory areas west of the International Date Line.

By 1941, Japan was determined to wage war. On November 10, V.Adm. Chuichi Naguma, placed in charge of the initial attack, issued his first operation order on the mission. The Striking Force of *Akagi, Kaga, Soryu, Hiryu, Shokaku* and *Zuikaku*, as well as other capital ships, sortied from Kure navy base between November 10 and 18, rendezvousing on the 22nd in Tankan Bay in the Kuriles. Adm. Yamamoto ordered the force to sortie on November 26. On December 2, he broadcast a prearranged signal that would launch the attack on Pearl Harbor: Niitaka Yama Nobore ("Climb Mount Niitaka"). Five days later, December 7, the Japanese Navy launched its surprise attack by aircraft, launched from carriers, at Pearl Harbor and the Philippines. The next day, the United States and Japan were officially at war.

Hiryu, Sister ship to the Soryu had her island on the port side, as did Akagi, the only two IJN aircraft carriers so configured. Hiryu was heavier by some 500 tons. Exact displacements of IJN carriers were difficult to pin point due to the burning of official records at the end of WW II. Source: www.history.navy.mil

Chapter 7

The Evolution of Aircraft Carriers

USS Intrepid operating in the Philippine Sea – November 1944. Source: www.history.navy.mil

Chapter 7 - The Early Attack Carriers

'We have hit the Japanese very hard in the Solomon Islands. We have probably broken the backbone of the power of their Fleet. They have still too many aircraft carriers to suit me, but soon we may well sink some more of them. We are going to press our advantages in the Southwest pacific and I am sure that we are destroying far more Japanese airplanes and sinking far more of their ships than they can build.'—*Franklin* D. Roosevelt, President of United States, 1942.

At the outbreak of World War II, the United States had in commission seven aircraft carriers and one escort carrier. USS *Langley*, the experimental ship officially classed as CV-1, had been assigned to duty as a seaplane tender on September 15, 1936.

After the abrogation by Japan from disarmament treaties, the U.S. took a realistic look at its naval strength. By Act of Congress on May 17, 1938, an increase of 40,000 tons in aircraft carriers was authorized. This permitted the building of USS *Hornet* (CV-8) and USS *Essex* (CV-9). On June 14, 1940, another increase in tonnage was authorized. Among the ships built under this program were the *Intrepid* and the new *Yorktown*. On July 19, an additional 200,000 tons for carriers was authorized.

Chapter 7

The Evolution of Aircraft Carriers

Adm. H.R. Stark, then Chief of Naval Operations, reported to the Secretary of the Navy: "In June 1940, the Congress granted the Navy an 11% increase in combat strength and, in July, a further increase of approximately 70%. When these ships and aircraft are completed, the U.S. Navy in underage and overage ships will include 32 battleships, 18 aircraft carriers, 91 cruisers, 325 destroyers, 185 submarines, and 15,000 airplanes.

"From 1921 to 1933, the United States tried the experiment of disarmament in fact and by example. This experiment failed. It cost us dearly in relative naval strength—but the greatest loss is TIME. Dollars cannot buy yesterday. Our present Fleet is strong, but it is not strong enough."

Additional tonnage was authorized December 23, 1941 and July 9, 1942. USS *Essex* (CV-9) was first of a series of early attack aircraft carriers of World War II.

CV-9 was to be the prototype of an especially designed 27,000-ton (standard displacement) aircraft carrier, considerably larger than the *Enterprise* and smaller than the *Saratoga*. These were to become known as the *Essex* class carrier, although this classification was dropped in the '50's.

On September 9, 1940, eight more of these carriers were ordered and were to become the *Hornet, Franklin, Ticonderoga, Randolph, Lexington, Bunker Hill, Wasp* and *Hancock*, CV-12 through -19, respectively. Reuse of the *Lexington, Wasp* and *Hornet* names was in line with the Navy's intent to carry on the traditions of the fighting predecessors: *Lexington* (CV-2) was lost in the Battle of the Coral Sea in May 1942; *Wasp* (CV-7) was sunk September that year in the South Pacific while escorting a troop convoy to Guadalcanal; *Hornet* (CV-8) was lost the following month in the Battle of Santa Cruz Islands.

USS Yorktown (CV-10) operating at sea in mid-1943, with a TBF "Avenger" flying over her bow. This photo was taken during carrier trials for the SB2C "Helldiver". Most of the planes on her flight deck are of that type. Source: www.history.navy.mil

Chapter 7

The Evolution of Aircraft Carriers

USS Randolph (CV-15) was the 13th Essex class carrier to be commissioned. She was the first of these carriers to enter combat without returning to the ship builder for post-shakedown work. She participated in the Iwo Jima, Okinawa, and Third Fleet operations against Japan in 1945. Photographed here alongside a repair ship at Ulithi Atoll, Caroline Islands, 13 March of the same year showing damage to her after flight deck resulting from a "Kamikaze" strike on 11 March. Photographed from a USS Miami (CL-89) floatplane Source: www.history.navy.mil

It is appropriate to comment here that the ships' names at commissioning date did not all bear the same name at the date of their programming. Names were changed during construction. *Hornet* (CV-12) was originally *Kearsarge*, *Ticonderoga* (CV-14) was the *Hancock*, *Lexington* (CV-16) was *Cabot*, *Wasp* (CV-18) was *Oriskany*, and *Hancock* (CV-19) was originally *Ticonderoga*.

Last two of the 13 originally programmed CV-9 class aircraft carriers, *Bennington* (CV-20) and *Boxer* (CV-21), were ordered on Dec. 15, 1941.

In drawing up the preliminary design for USS *Essex*, particular attention was directed at the size of both her flight and hangar decks. Aircraft design had come a long way from the comparatively light planes used in carriers during the Thirties. Flight decks now required more takeoff space for the heavier fighters and bombers being developed. Most of the first-line carriers of the pre-war years were equipped with flush deck catapults, but owing to the speed and size of these ships very little catapulting was done—except for experimental purposes. With the advent of war, airplane weights began to go up as armor and armament got heavier; crew size aboard the planes also increased. It was inevitable, noted

Chapter 7

The Evolution of Aircraft Carriers

the Bureau of Aeronautics toward the war's end in 1945, that catapult launchings would become more common under these circumstances. Some carrier commanding officers reported that as much as 40 percent of launchings were affected by the ships' catapults.

USS Bennington (CV-20) underway during her shakedown cruise in the western Atlantic or Caribbean area, 20 October 1944; She is painted in camouflage Measure 32, Design 17a (#1). Note the tanker at the top left in the distance. Source: www.history.navy.mil

The hangar area design came in for many conferences between Bureaus and much more official correspondence. Not only were the supporting structures to the flight deck to carry the increased weight of the landing and parked aircraft, but they were to have sufficient strength to support the tricing up of spare fuselages and parts (50 per cent of each plane type aboard) under the flight deck and still provide adequate working space for the men using the area below.

"At present," noted the Bureau of Construction and Repair in April 1940, "it appears that a few of the smaller fuselages can be triced up overhead in locations where encroachment on head-room is acceptable, and that the larger fuselages will have to be stowed on deck in the after end of the hangar. The number to be stowed will depend upon the amount of reduction in operating space in the hangar which can be accepted."

Capt. Marc A. Mitscher, then Assistant Chief BuAer, answered: "The question of spare airplanes is now under reconsideration in correspondence with the Fleet and the results decided upon will have a bearing in the case of CV-9."

Chapter 7

The Evolution of Aircraft Carriers

A startling innovation in CV-9 was a port side deck edge elevator in addition to two inboard elevators. Earlier, BuShips experimented with a ramp arrangement between the hangar and flight decks, up which aircraft were hauled by crane. This proved too slow. BuShips and the Chief Engineer of A.B.C. Elevator Co. designed the engine for the side elevator. Essentially, it was a standard elevator, 60 feet by 34 in platform surface, which travelled vertically on the port side of the ship. Capt. Donald B. Duncan, *Essex*'s first commanding officer, was enthusiastic. After the first four months of operation after commissioning, he wrote to BuShips:

"The elevator has functioned most satisfactorily in all respects and it is desired to point out some of the operational advantages realized with this type of elevator.

"Since there is no large hole in the flight deck when the elevator is in the 'down' position, it is easier to continue normal operations on deck, irrespective of the position of the elevator.

Curtiss SB2C-3 "Helldiver" aircraft bank over the carrier USS Hornet (CV_12) before landing, following strikes on Japanese shipping in the China Sea, circa mid-January 1945. Photographed by Lieutenant Commander Charles Kerlee, USNR - Source: www.history.navy.mil

The elevator increases the effective deck space when it is in the 'up' position by providing additional parking room outside the normal contours of the flight deck, and increases the effective area on the hangar deck by the absence of

Chapter 7
The Evolution of Aircraft Carriers

elevator pits." The elevator performed well, its machinery less complex than the two inboard elevators, requiring about 20 per cent fewer man-hours of maintenance. Capt. Duncan recommended that consideration be given using two deck edge elevators, one on each side. BuShips, in forwarding the recommendation to BuShips, offered another advantage for consideration: a conventional elevator suffering a casualty while in the "down" position "would leave a large hole in the flight deck while the deck edge type would cause only minor and non-critical loss of flight deck area."

BuShips, obviously pleased with the operational performance of the new elevator—the first of its kind—reluctantly turned down the recommendation, however. The Bureau noted that the addition of a starboard deck-edge elevator would not permit an *Essex* class aircraft carrier to transit the Panama Canal. Any other location for a second such elevator would involve structural and arrangement changes too extensive to be considered.

USS Yorktown (CV-10) hanger deck with munitions mates arming their bombs; flight crews are in the background in discussion about the day's mission. The Yorktown was the third Essex class carrier commissioned, sponsored by Mrs. F.D. Roosevelt. Source: www.history.navy.mil

On April 28, 1941, keel for the USS *Essex* was laid at Newport News Shipbuilding and Dry Dock Co. On October 2, the following year, her prospective commanding officer filed his first weekly progress and readiness report to the Chief of Naval Operations (CNO). He noted that there was marked

Chapter 7

The Evolution of Aircraft Carriers

speed-up of work on the ship during the preceding month and estimated that the ship would probably be delivered on February 1, 1942.

"There are certain items that have been authorized for installation on the CV-9 - 19 class carrier," he said, "but will not be accomplished on this vessel prior to delivery." The ship was launched July 31, 1942.

RAdm. Walter S. Anderson, president of the dock trials and inspection team of CV-9 on December 23, 1942, noted a few of these discrepancies in his report:

"Due to late authorization of a number of changes arising out of recent war experiences, the volume of uncompleted hull work was greater than normal. The Board regrets that the catapults for this vessel were not delivered in time for installation, as military value of the vessel would be much improved thereby. Only the starboard flight deck track was installed. This class of carriers is designed to include cruising turbines as part of the main drive turbine installation. However, due to production difficulties and as a result of efforts to expedite delivery, cruising turbines were omitted. Space and connections for their future installation are provided and this can be accomplished with very little alteration."

Nevertheless, the Board was pleased and impressed with progress on construction of the *Essex*. Adm. Anderson recommended acceptance of the ship. "On 31 December 1942," he said, "only slightly over 20 months will have elapsed since keel-laying, which is, in the opinion of the Board, a record worthy of commendation. This indicates a high degree of cooperation between the Supervisor of Shipbuilding, the Newport News Shipbuilding and Dry Dock Co., and representatives of the officers and men of the ship's company." On the last day of 1942, USS *Essex* was commissioned.

Capt. Duncan was proud of his new command, but not so impressed as to ignore certain discrepancies that still existed. The ventilation system, for instance, was less than satisfactory. BuShips sent representatives to the ship to assist in correcting discrepancies, during sea trials March 1 in the North Atlantic and, a month and a half later, when the ship was again at Norfolk and still had complaints.

As other CV-9 carriers were launched, the complaints continued to be registered. BuShips investigated the ventilation system as installed in USS *Intrepid* (CV-11) and outlined corrective measures in future carriers of the class.

Requested to comment on the adequacy and operation of the trash burner installed in the *Essex*, Capt. Duncan started off quietly enough. "It is most unsatisfactory," he said. Then he warmed to his subject. "It is doubtful if it could be worse. It is in the very center of the office spaces. There is no satisfactory place for collection of trash waiting its turn to be burned. All of it has to be carried through the passageways in the vicinity of the departmental offices. The heat from the trash burner when it is operating (which is not often because it is

Chapter 7

The Evolution of Aircraft Carriers

usually broken down) is such as to make the surrounding spaces almost untenable.

"The design of the trash burner is poor. Its construction is worse. The ship had not been in commission a month before it practically fell apart. The brick work fell down, the door fell off and it suffered other casualties too numerous to mention. It has taken constant attention from the Engineer's force to keep it operating at all and the heat generated in the compartment in which it is located is such that it is physically impossible for men to stay in it for continuous operation." - The trash burner was redesigned.

Lexington was commissioned on February 17, 1943, followed by *Yorktown* on April 15, *Bunker Hill* on May 25, *Intrepid* on August 16, *Wasp* on November 24, and *Hornet* on November 29 that year. In 1944, *Franklin* was commissioned on January 31, *Hancock* on April 15, *Ticonderoga* on May 8, *Bennington* on August 6, and *Randolph* on October 9. The last of the programmed 13 CV-9's, *Boxer*, was commissioned on April 16, 1945.

The lighting system installed in the *Lexington* came under the scrutiny of BuShips. Generally, it was considered inadequate—"in intensity and quality"—in many passageways and compartments, in addition to the running, signal, and anchor lights. A survey of the system produced the following action on the outside lights: the ahead masthead light was relocated to the forward edge of the foretruck (frame 92), the ahead range light was moved forward and shielded from illuminating the deck below, the astern masthead light was moved higher, so as not to interfere with gunnery, and the astern range light was removed.

Nineteen more *Essex*-class ships were ordered or scheduled, starting with ten of them on August 7, 1942. They were *Bon Homme Richard* (CV-31) *Kearsarge*

USS Independence (CVL-22) In San Francisco Bay, CA, 15 July 1943, the day her hull number was changed from CV-22 to CVL-22. She has nine SBD scout bombers amidships and nine TBM torpedo planes amidships and forward Source: www.history.navy.mil

Chapter 7

The Evolution of Aircraft Carriers

(CV-33), *Oriskany* (CV-34), *Reprisal* (CV-35), *Antietam* (CV-36), *Princeton* (CV-37), *Shangri La* (CV-38), *Lake Champlain* (CV-39), *Tarawa* (CV-40), and *Crown Point* (CV-32) - later renamed Leyte. The last three ordered were *Valley Forge* (CV-45), *Iwo Jima* (CV-46), and *Philippine Sea* (CV-47). The keels were laid for Reprisal and Iwo Jima on July 1, 1944 and January 29, 1945, but both were cancelled on August 11, 1945. Six additional 27,000-tonners, CV's 50 through 55, were canceled on March 27, 1945.

USS Cowpens (CVL-25) was one of nine cruiser-to-aircraft carrier conversions.
Source: www.history.navy.mil

In recap, after WW II erupted and until its successful conclusion by Allied forces, the U.S. Navy ordered 32 aircraft carriers of the CV-9 class, of which the keels of 25 were laid down. A total of 17 were actually commissioned during the war years. The total number of CV-9's commissioned—including those commissioned after the war—was 24.

Several characteristics marked the *Essex* class carriers upon their introduction to the Fleet. The pyramidal island structure, for instance, rose cleanly from the starboard side, topped by a short stack and a light tripod mast. The port elevator was also a distinguishing feature, along with the two inboard elevators. *Ticonderoga, Randolph, Hancock, Bennington* and *Boxer*, as well as hull numbers from CV-31 on, had rounded bows extending beyond the flight deck.

Overall lengths varied within this class; they were either 872 feet long or 888 feet. It is interesting to note that they had a uniform water line length of 820 feet. All were armed with 12 five-inch .38 caliber dual purpose guns, but some had 17 quadruple 40mm anti-aircraft mounts while others had 18. They also had an extensive 20mm AA armament. Generally, there were accommodations aboard each for 360 officers and 3088 enlisted men.

Chapter 7

The Evolution of Aircraft Carriers

Except for CV-2 and CV-3, *Lexington* and *Saratoga*, the power plants were increased over other aircraft carriers in the Fleet. The machinery was "entirely modern in design and arranged so as to gain the maximum resistance to derangement and battle damage. There are eight control superheat boilers arranged in four firerooms. Steam lines are such that the boilers in each fireroom can be connected to one main machinery unit so that the plant can be operated as four separate units." They had four screws.

These carriers had better protecting armor than their predecessors (again excepting *Lexington* and *Saratoga*), better facilities for handling ammunition, safer and greater fueling capacity, and more effective damage control equipment.

The Tactical employment of U.S. carriers changed as the war progressed. In early operations, through 1942, the doctrine was to operate singly or in pairs, joining together for the offense and separating when on the defense—the theory being that a separation of carriers under attack not only provided a protective screen for each, but also dispersed the targets and divided the enemy's attack. Combat experience in those early operations did not bear out the theory and new proposals for tactical deployment were the subject of much discussion. As the new *Essex* and Independence class carriers became available, these new ideas were put to the test.

The *Independence* class carriers—light carriers, designated CVL's—were products of an effort to increase this country's sea-going air strength in the early days of the war. Nine keels to light cruisers of the *Cleveland* class were laid down at the New York Shipbuilding Corp. yard at Camden, N. J., three of them before the war started. They were to have been the *Amsterdam* (CL-59), *Tallahassee* (CL-61), *New Haven* (CL-76), *Huntington* (CL-77), *Dayton* (CL-78), *Fargo* (CL-85), *Wilmington* (CL-79), *Buffalo* (CL-99), and the *Newark* (CL-100). They eventually became the *Independence, Princeton, Belleau Wood, Cowpens, Monterey, Langley, Cabot, Bataan*, and the *San Jacinto*, CVL's 22 through 30, respectively.

Launch of the USS Princeton down greased skids at the New York Ship Yard October 1942.
Source: www.history.navy.mil

Chapter 7

The Evolution of Aircraft Carriers

Naming and designating these last four ships sometimes went through a rigorous and confusing metamorphosis, Neither *Cabot* nor did *Bataan* encounter any difficulty. The names and designations were reached in June and July 1943 without attending problems But *Fargo* was named *Crown Point* (CV-27) when the decision was reached to convert her to an aircraft carrier. Then, on July 15, 1943, her name was changed to USS *Langley* and she was given the designation CVL. (Actually, all these cruiser-to-carrier conversions were originally designated CV's when the decision to convert was made; all were re-designated CVL's on the same day.)

Fast Carrier task forces included both Essex and Independence class carriers, shown above is an official declassified photograph of the USS Independence off the Mare Island Navy Yard - 13 July 1943.
Source: www.history.navy.mil

The *Newark* (CL-100) had a rougher time of it. On June 2, 1942, she was changed to CV-30; on June 23, her name was changed to *Reprisal*, which she kept for a little over six months. On Jan. 6, 1943, her name was again changed, to *San Jacinto*.

Chapter 7

The Evolution of Aircraft Carriers

The light carriers displaced 11,000 tons standard. In design, the bridge was box-like in appearance, with a small crane forward. They had four stacks, paired off in twos, on the starboard side, aft of the island. These stacks angled out from the hangar deck and rose vertically above the flight deck level.

USS Shangri-La underway in the Pacific, 17 August 1946; use of her flight deck is flourishing with her crew standing in divisions. For some reason, the ship's deck pennant number has been replaced by a Z. Source: www.history.navy.mil

As the *Essex* and *Independence* class carriers joined the Fleet in increasing numbers, it was possible to operate several carriers together, on a continuing basis, forming a carrier task group. Tactics changed. Experience taught the wisdom of combined strength. Under attack, the combined anti-aircraft fire of the task group carriers and their screen provided a more effective umbrella of protection against marauding enemy aircraft than was possible when the carriers separated. When two or more of these task groups supported each other, they constituted a fast carrier task force.

The first attempt to operate a multi-carrier group occurred on August 31, 1943, during a raid on the Japanese-held island of Marcus. Task Force 15, which conducted the raid, consisted of *Yorktown* (CV-10), *Essex* (CV-9) and *Independence* (CVL-22), the cruisers *Nashville* and *Mobile*, the battleship *Indiana*, and ten destroyers. Aircraft were launched from the carriers at a point approximately 130 miles north of the island.

Chapter 7

The Evolution of Aircraft Carriers

On October 5th and 6, 1943, Rear Admiral Alfred E. Montgomery led Task Force 14 on a second raid on Wake Island. The task force was comprised of two task groups, operating a total of six aircraft carriers— *Essex, Yorktown* (CV-10), *Lexington* (CV-16), *Independence, Belleau Wood,* and *Cowpens* —seven cruisers and 24 DD's, the largest carrier task force yet assembled. In the course of the two-day strikes, ship handling techniques for a multicarrier force, devised by RAdm. Frederick C. Sherman's staff was tested under combat conditions.

Adm. Chester W. Nimitz, then Commander in Chief, Pacific Fleet, dispatched his congratulations. "The thorough job done on Wake by planes and ships of your task force will have results reaching far beyond the heavy damage inflicted."

The words were prophetic. Lessons learned from operating the carriers as a single group of six, as two groups of three and three groups of two, provided the basis for many tactics which later characterized carrier task force operations. With the evolution of the fast carrier task force and its successful employment in future operations, the rising sun of the east began slowly to sink in the west.

Chapter 8

The Evolution of Aircraft Carriers

TBM "Avenger" torpedo plane is catapulted from USS Makin Island (CVE-93), circa 1945. Collection of VAdm, Calvin T. Durgin. Donated by his daughter, Mrs. Phyllis Durgin Sherrill, 1969. The CVE class of Escort Carrier were seldom photgraphed. Source: www.history.navy.mil

Chapter 8 - Emergence of the Escort Carriers

'The story of the escort aircraft carriers is like a story with a surprise ending. When the United States began to build them, there was a definite purpose in view—fighting off submarines and escorting convoys. But as the war progressed, the small carrier demonstrated surprising versatility. It became a great deal more than its name implies. From a purely defensive measure, the escort carrier emerged as an offensive weapon.'
— FAdm. Chester W. Nimitz, USN, CinCPacFlt / CinCPOA, 1945

Toward The End of World War I, Great Britain experimented in converting light cruisers to airplane carriers—notably in HMS *Cavendish* of 32 knots and about 10,000 tons displacement. But with the signing of the Armistice, the project was abandoned. Despite this, it was a subject of interest in the following years.

In 1925, the General Board seriously considered the conversion of cruiser hulls to aircraft carriers. Although treaty limitations restricted the building up of carrier strength, there was sufficient uncommitted construction tonnage to permit the building of more carriers than the U.S. Fleet had. Could this

Chapter 8

The Evolution of Aircraft Carriers

uncommitted tonnage be best employed in building small carriers? The Board's answer can best be summed up in this excerpt from its report:

"Incomplete studies of the subject by the Bureau of Construction and Repair and the meager information available concerning the performance of airplanes from carriers of approximately 10,000 tons displacement does not justify building them at this time."

But the subject of "light" carriers was of recurrent interest to the U.S. Fleet. In May 1927, LCdr. Bruce G. Leighton prepared a paper in which he analyzed the problem. He titled it, "Light Aircraft Carriers, A Study of their Possible Uses in So-Called 'Cruiser Operations,' Comparison with Light Cruisers as Fleet Units." Though the title may have been cumbersome, the document was impressive. He forecast every fundamental combat requirement of the later-day CVL's and CVE's, including the bombing of capital ships, support of fleet operations, anti-submarine work, scouting and reconnaissance, and the reduction of enemy shore bases. He concluded that "all things considered, it might well be considered as a worthy substitute for the light cruiser, or even distinctly preferable to the cruiser."

For the next dozen years, the subject interjected itself spasmodically and unsuccessfully into Navy thinking. But in March 1939, Capt. John S. McCain, Sr., then commanding the *Ranger*, wrote to the Secretary of the Navy advocating the building of at least eight "pocket-size" carriers of cruiser speed. These were not meant to replace the CV's, but to supplement them, giving force commanders much more flexibility in the use of ship-based aircraft at sea, without jeopardizing the much more costly heavy carriers. RAdm, Ernest J. King, in his endorsement to the letter, was not at all enthusiastic about this scheme. He suggested that existing aircraft carriers carry the maximum number of planes permissible as a better solution than the construction of smaller carriers.

The matter was not entirely dropped, however, for the Bureau of Construction and Repair was considering, and even drawing plans for the conversion of 20 or 21-knot passenger ships, creating experimental carriers with short flight decks. By November 1940, the Chief of Naval Operations brought these considerations to an abrupt halt, basing his decision on a letter from SecNav to the Chairman of the U.S. Maritime Commission. SecNav wrote:

"The characteristics of aircraft have changed, placing more exacting demands upon the carrier. These demands are such that a converted merchant vessel will no longer make as satisfactory an aircraft carrier as was the case when the plans for those vessels were being drawn."

In commenting on the beginning of escort carriers, historian Lt. William G. Land, USNR (Functional Development of the Small Carrier [CVE]) says, "The escort carrier was forced upon the Navy by the President."

Chapter 8

The Evolution of Aircraft Carriers

Indeed, President Franklin D. Roosevelt did actively enter the "light" carrier controversy. Great Britain had been at war with Germany since September 1939. Since that time and before the U.S. entered the war, large numbers of U.S.-built military aircraft were sold to the British. The U.S. had need for an aircraft-carrying ship to speed delivery. By mid-February 1941, RAdm, W. F. Halsey (later Fleet Admiral) had written to Commander-in-Chief, U.S. Fleet:

"A previously stated expectation that the Navy would be called upon to provide transport for Army aircraft, has now materialized in the current diversion of *Enterprise* and *Lexington* to transport 80 pursuit planes from the West Coast to Hawaii. To continue with primary reliance on aircraft carriers for such work, as is our present necessity, seriously endangers the availability of air-offensive power in the Fleet."

Adm. Husband E. Kimmel, in endorsing this letter from his Commander Aircraft Battle Force to the Chief of Naval Operations, fully concurred and pointed out that on five separate occasions in the past he had himself urged such action.

Earlier, on October 21, 1940, CNO had received a memorandum from the President's Naval Aide advising him that President Roosevelt proposed the Navy acquire a merchant ship and convert it to an aircraft carrier, accommodating 8 to 12 helicopters (not yet operated by the Navy) or airplanes capable of landing or taking off in a small space. The purpose of this type carrier was to "provide quick conversions for carrying small planes which could hover ahead of convoys detect submarines and drop smoke bombs to indicate their locations to an attacking surface escort craft."

Typical of the later CVE class aircraft carriers, the USS Bairoko had much cleaner lines than those early CVE class light carriers which were frantic conversions prior to the U.S. entering WW II. Source: www.history.navy.mil

Chapter 8

The Evolution of Aircraft Carriers

CNO decided on the last day of 1940 that the Chairman of the Maritime Commission would be consulted to determine the availability of ships for this purpose. On January 2nd, it was found that two Danish ships might permit conversion, but later investigation proved this would not be possible. The results of this January 2 conference determined that the ships (two— one was to be sold to Great Britain) selected "should be of the same or very similar design in order that the plans made for one could be applicable to both; that the airplanes should be further investigated to determine the type and availability; that an armament of four AA pom-poms and one 5" surface gun should be such as to insure stability at all stages of loading." These converted merchant ships were to fill the need later expressed by Adm. Halsey, the transport of aircraft, as well as to provide protection to Allied convoys.

On January 6, 1941, Adm. Harold R. Stark, CNO, convened a conference in his Washington office to discuss merchant-conversion. The autogiro type aircraft was considered of dubious usefulness because of its inability to carry any load other than smoke bombs; an aircraft, to meet the purpose designed, must have some offensive characteristics. An abbreviated deck was ruled out. The converted ship should be diesel-driven in order to eliminate smokestacks. The decision was made to obtain from the Maritime Commission, if possible, C-3 cargo ships.

On the following day, CNO was informed that two diesel-driven C-3 type ships, the *Mormacmail* and *Mormacland* would be suitable for conversion and were available. He was told by President Roosevelt that any plan which would take more than about three months to complete conversion would be unacceptable. This, in effect, placed pressure on the project. The idea of substituting "blimps" for autogiros or heavier-than-air craft was flirted with but, by January 15, was "out of the picture."

The *Mormacmail* was acquired on March 6, 1941. On June 2—just within the three-month limitation set by the President—she emerged from conversion and was placed in commission as the aircraft escort vessel USS *Long Island* (AVG-1), commanded by Cdr. Donald B. Duncan who, on December 31, 1942, was to be the first commanding officer of USS *Essex*.

Early plans for the conversion called for the installation of a 305-foot flight deck on the *Mormacmail*, but the Bureau of Aeronautics required at least 350 feet to safely land SOC Sea Gulls aboard. Upon commissioning, *Long Island* had a deck length of 362 feet. She had one elevator, handled 16 planes, had a trial run speed of 17.6 knots, and berthed 190 officers and 780 men.

The *Mormacland*, acquired at the same time, was similarly converted and was turned over to the British as HMS *Archer* (BAVG-1) when it was completed the following November. Experience with the BAVG and the two British conversions led the British to believe that the diesel-driven ships were too slow

Chapter 8

The Evolution of Aircraft Carriers

for their purpose as special escort vessels—although they were no slower than the later *Bogue* class escort carriers.

Long Island was used primarily as a training ship during the remaining peacetime months of 1941. She was subjected to tests and experiments— much the way USS *Langley* had been in her early days—to obtain data needed for the construction of later escort carriers. As a result of the Navy's experiences with this ship, other CVE's were outfitted with two elevators instead of one, the flight decks were lengthened, and the antiaircraft power was increased.

On December 26, 1941, SecNav approved the conversion of 24 merchant hulls for the 1942 shipbuilding program and, in March, ordered the conversion of cruiser hulls which became the CVL's. Cdr. Leighton's 1927 paper was proving its farsightedness.

Naval Aviation historian, Dr. Henry Dater, traced the next developments in a paper published in Military Affairs:

"There were only 20 C-3 hulls available for conversion, ten of which were earmarked for the Royal Navy and ten for the United States. The new ships were improved by the substitution of a steam turbine power plant for the diesel engines employed in the *Long Island* and *Charger* [the latter was re-designated CVE-30 and replaced CVE-1 as a training ship when the *Long Island* was pressed into service, ferrying planes and pilots at the outbreak of war], and by the addition of a slightly larger flight deck (436 by 79 feet), a small island, and a considerably larger hangar space."

"They were referred to either as the CVE-6 class, from the numerical designation of HMS *Battler*, or as the *Bogue* class, from the first ship to operate with the U.S. Navy."

"The remaining four CVE's authorized for the 1942 program were converted from Cimarron class fast fleet oilers and were known as the *Sangamon* (CVE-26) class. These were considerably larger, having a flight deck of 503 feet by 85 feet, and were able to accommodate two small squadrons of aircraft. Because of their size, work was rushed on them during the summer of 1942 so that they would be available for the North African invasion in the autumn."

Before the U.S. entered the war, German U-boats hovered near British coastal ports and picked off merchant ships with ease. Land-based RAF planes drove the German submarines further out to sea. To make matters more difficult for the enemy, convoys sailed closer together, opening up larger areas of the North Atlantic for the German subs to search. The Germans solved this problem by developing the "wolf pack" technique of operating in groups, then concentrating for the kill when convoys were sighted.

"It was this technique which created the British desire for aircraft escort vessels in late 1940 and 1941," wrote Dr. Dater. "With the entry of the

Chapter 8

The Evolution of Aircraft Carriers

United States into the conflict the Germans found easy picking off the American coast, but it was only a matter of time until land-based air on this side of the Atlantic drove them out to sea once more. There in mid-ocean was a vast area in which the convoys did not have the assistance of aircraft. By early 1943 it became evident that the decisive campaign was to be fought in that area."

The air officer of the *Bogue* described escort procedures during March and April 1943:

"The ship was stationed inside the convoy for this work. The convoys were in columns of five ships each with about 700 yards between columns. They left a double space in the middle in the center of which they placed the *Bogue*. The other escorts were placed around the convoy in a half circle. The idea was, if possible, to use our catapults and to stay in our center position when launching our planes so there wouldn't be any wide separation. As it happened, we had westerly winds on the East-bound convoy so we had to turn around to launch planes and to take them aboard. Consequently, the separation was fairly large due to the fact that it was what is called a high speed convoy, 'nine knots!'"

Bogue Class escort carriers were products of the 1942 shipbuilding program. They were converted from Maritime Commissioned C-3 hulls as quickly as possible. Source: www.history.navy.mil

Though this tactic met with considerable success at first, it was primarily defensive. A new technique was found more effective. A small task group took up a position where it could throw its support to either of two convoys in a general area. Escort carrier-based aircraft scouted ahead, searching out German U-boats before the submarines could make contact. This permitted the carriers to be released from the difficult maneuver necessary in the central slot of the

Chapter 8

The Evolution of Aircraft Carriers

convoy. Out of this technique emerged the successful hunter-killer tactic that eventually freed Allied shipping in the North Atlantic.

The *Sangamon* class escort carriers, built as fleet oilers under the Merchant Marine Act of 1936, were completed in 1939, but in the 1942 shipbuilding program were slated for reconfiguration to aircraft carrier characteristics. Only four hulls were on hand. "Had more oiler hulls been available," wrote Lt. Land, "they would have become the prototype of the small carrier for the ensuing year's program. But the overwhelming need for fleet oilers—to make possible our logistic advance—prevented this type of hull from being again used for carriers, until 1944."

The *Sangamons* had an over-all length of 553 feet, a speed of 18.3 knots, a trial displacement of 23,235 tons, and carried 120 officers and 960 men. They were armed with two five-inch, 38 caliber guns, two quad and ten twin 40mm AA mounts. They were equipped with two hydraulic catapults forward.

USS Suwanee was one of four escort carriers converted from Cimmaron class fleet oilers. They were rushed to completion for battle duty before WW II. Source: www.history.navy.mil

"With the CV's, except *Ranger*, being employed in the Pacific," wrote historian Land, "planning for the North African landings depended on the completion of the AO conversions of *Suwannee*, *Sangamon*, *Chenango*, and *Santee*. For this reason, *Suwannee* had to cut down on its pre-commissioning period, fitting out, and shakedown in order to be substituted in the final plans for the much smaller *Charger*, the ex-BAVG which had been doing regular duty as qualification carrier in Chesapeake Bay. *Santee*,

Chapter 8

The Evolution of Aircraft Carriers

likewise, was barely completed in its essentials and had had hardly any exercise with its air group before its first combat operation was to begin."

Capt. William D. Sample, commanding *Santee*, wrote of the hectic early days aboard:

"*Santee* left Norfolk Navy Yard 13 September 1942 with Yard workmen still on board and her decks piled high with stores. During that first month, the *Santee* returned to the yard twice and was never free of the Yard workmen. The completion of the ship continued while the fitting out and shakedown were proceeding together. At the end of the month, the air group had operated aboard only a day and a half and guns had been fired only for structural tests."

"The Navy Yard had done an almost impossible task in getting the ship out in time for the pending operations but, in so doing, only the essentials had been completed, and it was then necessary for the ship to install, adjust, calibrate and repair until the ship could use her battery and equipment. The service experience necessary to test many of the questionable features of the ship's design was soon obtained in a wintry gale encountered en route to Bermuda. The two forward boats were carried away, the new upper decks proved to be sieves and the repair work of the ship's force got underway in earnest."

USS Casablanca (CVE-55) was the first of 50 escort carriers mass-produced by Kaiser Shipyard. It's quite easy to tell the difference between a conversion and a keel up carrier even when done as quickly as possible like the Casablanca was. She had fine lines for her class
Source: www.Zap16.com

Chapter 8

The Evolution of Aircraft Carriers

The carrier *Chenango* was used, in the North African operation that followed, as a ferry carrier for Army P-40's on the outward trip, as a fuel supply ship while moored at Casablanca, and as a fleet escort—with a borrowed air group furnishing air cover-on the return trip.

Her sister ships, however, launched TBF-1 *Avengers*, SBD-3 *Dauntless* and F4F-4 Wildcat aircraft in support of landing operations for the capture of Casablanca and Port Lyautey. They were units of Task Force 34. As part of the Northern Attack Group, *Sangamon* and *Chenango* assisted troops landing at Mehedia, not far from Port Lyautey. *Ranger* and *Suwanee* provided air cover for the Center Attack Group at Fedhala, northeast of Casablanca. *Santee* was the only carrier assigned to the Southern Attack Group, providing combat air patrol and anti-submarine patrol for the landing force at Safi—the only port in Morocco, other than Casablanca, that would permit the landing of 28-ton General Sherman tanks. It was for the capture of Casablanca that these tanks were needed.

Between November 8th and 11th, 1942, *Suwanee* launched 255 combat sorties; Santee, 144, and *Sangamon*, 183.

During *Sangamon*'s participation in the Northern Attack Group operation, her planes were called upon to neutralize a Kasba or citadel, which guarded the Port Lyautey airdrome. Several SBD's delivered bombs on target. "The garrison then," wrote Samuel Eliot Morison, "came out with their hands up and our infantry walked in." BY November 15, *Sangamon*'s part in the invasion of North Africa was completed and she sailed for Hampton Roads.

Planes in the *Suwanee* joined those based in the *Ranger* in bombing missions during the Battle of Casablanca. The *Suwanee*, like the *Santee* at Safi, encountered light winds. Many landings were made aboard with only 22knot winds across the deck.

Despite the greenness of the crews in the *Sangamon*, generally, they gave a good account of themselves. Commented CinCLant: "The CVE's proved to be a valuable addition to the Fleet. They can handle a potent air group and, while their speed is insufficient, they can operate under most weather conditions and are very useful ships."

The U.S. Navy escort carrier USS Chenango (CVE-28) operating in the Pacific in 1944 painted in Camouflage Measure 33, Design 10A. She played many roles in WW II including oil transport, escort and launch platform for Army Air Corp aircraft. She successfully destroyed dozens of targets during her illustrious service to the United States Navy. Source: www.history.navy.mil

Chapter 8

The Evolution of Aircraft Carriers

Their missions in the invasion of North Africa completed, *Sangamon*, *Chenango*, and *Suwanee* were dispatched to the Pacific. By the end of 1942, U.S. carrier strength in the Pacific had been reduced to the *Enterprise* and the *Saratoga*.

In the meantime, President Roosevelt announced a need for more escort carriers. Shipbuilder Henry J. Kaiser had impressed the President with the merits of a plan which would permit the mass production of escort carriers, under a program to be supervised by the Maritime Commission.

The first of these, USS *Casablanca* (CVE-55), was commissioned July 8, 1943, and gave its name to the class - CVE-55 through CVE-104. They were also referred to as Kaiser Class escort carriers. The Kaiser yard completed its 50-ship program on July 8, 1944. This was an impressive achievement in wartime production program.

The *Casablanca* class had an over-all length of 512 feet, 3 inches, a speed of 19.3 knots, a trial displacement of 9570 tons, and carried 110 officers and 750 men. They had one five-inch, 38 caliber gun and eight twin 40mm AA mounts. The aircraft complement consisted of 12 Grumman built TBM "Avengers" and 16 General Motors built FM-2 "Wildcats"; in the flight deck was a single hydraulic catapult, forward.

A Kamikaze is photographed here at the time of impact as it explodes into the U.S. Navy escort carrier USS Suwannee (CVE-27) in the waters off the Philippines on 26 October 1944. This incredible photo was taken from the USS Sangamon (CVE-26). Source: www.history.navy.mil

Chapter 8

The Evolution of Aircraft Carriers

Final details were worked out for a new class escort carrier during the trials of the *Sangamon* and *Santee* and during the planning for the 1944 building program. These ships were the first Navy-designed escort carriers for which hull and propeller model tests were carried out at the David W. Taylor Model Basin. The design was formally approved by CNO on December 10, 1942 and the contract was let on January 23, 1943. The first of these carriers was the *Commencement Bay* (CVE-105) from which the class got its name. It had an over-all length of 557 feet, a speed of 19 knots, and a trial displacement of 23,100 tons. Few of these ships saw action in the war— the *Commencement Bay* was commissioned in November 1944. Only nine others were commissioned before V-J Day the following September. They incorporated all lessons learned since the *Long Island* was commissioned.

As the escort carriers gained experience, they earned the respect of the Fleet by proving themselves versatile in anti-submarine warfare. The *Sangamon* class first demonstrated combat capability in the support of the North African invasion. The first major carrier-supported amphibious landing in the Pacific was the capture of the Gilberts and Marshalls. Eight escort carriers participated, two of the *Bogue* class, three of the *Sangamon* class, and three of the *Casablanca* class. The changing status of these vessels is reflected in their re-designation. Originally identified as aircraft escort vessels (AVG's), they were re-designated on August 20, 1942, auxiliary aircraft carriers (ACV's), and finally, on July 15, 1943, a directive changed the escort carrier symbol to CVE, reclassifying them as combatant ships.

At the end of the North African invasion, RAdm, Calvin T. Durgin (then Capt.) evaluated the effectiveness of the escort carriers when he presented his report:

"Due to their low speed, lack of protection and light armament, it is considered hazardous to employ a CVE group in operation where there is likely to be an effective enemy opposition. Such a group can, however, be used to advantage, and is capable of inflicting substantial damage to the enemy in assault where the enemy air and sea opposition is negligible or when it is being contained by other superior forces. When this situation exists, the CVE is well equipped to provide all support until landing strips are established ashore, and it can be effectively employed for bombardment spotting, combat air patrols over beaches and surface forces, for all forms of air reconnaissance missions and for bombing, rocket and strafing attacks."

His experience with escort carriers was to stand him in good stead. On December 13, 1944, the functional type command, Escort Carrier Force, Pacific, was created; RAdm. Durgin was placed in command.

Chapter 8

The Evolution of Aircraft Carriers

*USS Barnes (CVE-20) with Grumman TBF "Avenger" aircraft parked on her flight deck, 8 January 1944
Source: www.history.navy.mil.*

The establishment of this force was made possible by the increasing number of carriers—notably of escort design—made available to the Fleet. Experience at Palau and Morotai and the difficulties encountered later at Leyte all pointed to the need for better planning in advance of operations if the CVE's were to perform efficiently their enlarged responsibilities. Adm. Durgin's command held administrative control over all escort carriers operating in the Pacific, except those assigned to training and transport duty.

On December 15, 1944, the escort carriers provided direct support for landings on Mindoro, and in the assault area on the next two days. Between January 3-22, 1945, 17 escort carriers covered the approach of the Luzon Attack Force against serious enemy air opposition from Kamikaze pilots. This force of ships, Task Group 77.4, conducted preliminary strikes in the assault area, covered the landings in Lingayen Gulf, and supported the inland advance of troops ashore.

In 1945 the CVE's saw a great deal of action. On the last three days of January, six escort carriers under RAdm. Sample (as Capt., first C.O. of *Santee*) provided air cover and support for landings by Army troops at San Antonio near Subic Bay, and at two other nearby Philippine beaches. In February, Adm. Durgin directed his carriers in the battle for the capture of Iwo Jima. In March, the Okinawa campaign began, the last, and, for naval forces, the most violent major amphibious campaign of World War II. As Task Group 52.1, Adm. Durgin, with an original strength of 18 escort carriers, conducted pre-assault strikes and

Chapter 8

The Evolution of Aircraft Carriers

supported the occupation of Kerama Retto, joined in the pre-assault strikes on Okinawa, and, from a fairly restricted operating area southeast of the island, supported the landings and flew daily close support for operations ashore until the island was secure on June 21.

The U.S. suffered few losses to the enemy in these ships. Five carriers of the

TBM Avenger caught up on the walkway and rail of the USS Long Island after a rather difficult landing; the shaken pilot is being helped from the aircraft by one of the crew members
Source: www.history.navy.mil

Casablanca class were lost in the Pacific; one *Bogue* class was torpedoed in the Atlantic. During the war years, 76 CVE's of various classes were commissioned, in addition to the *Long Island*, commissioned months earlier. Seven more *Commencement Bay* class carriers were commissioned during the post-war years. During the war, four sister ships to *Long Island* were transferred to the British, as were 34 additional escort carriers of the *Bogue* class. Four were sunk; at the end of the war, the rest were returned to the U.S. from Lend-Lease and were either sold or placed in the reserve fleet.

Through fulfilling a basic need of transporting large numbers of assembled aircraft to various theaters of war, the quickly conceived and executed escort carrier developed into an anti-submarine warfare weapon that defeated the German U-boat threat in the North Atlantic. They provided combat capability in the support of fleet operations in both the Atlantic and the Pacific. In short, they displayed versatility, proven under the pressures and urgencies of a war that engulfed the world.

Chapter 8

The Evolution of Aircraft Carriers

In a remarkable naval battle off the Philippine island of Samar on the 25th October 1944, an American force consisting of six CVEs, three destroyers and four destroyer escorts was attacked by a Japanese fleet consisting of four battleships, eight cruisers and eleven destroyers. In a running battle that lasted two hours, three American destroyers, the Hoel, Heerman and Johnston, and one destroyer escort, the Samuel B Roberts, repeatedly drove off the Japanese ships attempting to engage the carriers. One CVE, the Gambier Bay (shown above) was sunk by Japanese gunfire. In exchange, another CVE, the Kalinin Bay, used her single five-inch gun to engage the Japanese cruiser Chokai (ten eight inch and eight five inch guns) in a gunnery duel. Incredibly, the Kalinin Bay won the duel, damaging the cruiser so badly that she had to be scuttled. As the badly-mauled Japanese force retreated, Admiral Sprague heard a nearby sailor exclaim: "Damn it, boys, they're getting away!" Details of this naval battle can be found in the book "Last Stand of the Tin Can Sailors" by James D Hornfischer. Photo source : www.history.navy.mil

The Evolution of Aircraft Carriers

USS Midway (CVB-41) was the first of six planned carriers of a new design. Construction began during World War II. Toward the end of hostilities, three of the new carriers were cancelled. Upon delivery, the Midway class carriers were the mightiest aircraft carriers in the world.

This photograph was taken from the USS Philippine Sea (CV-47) during a gale east of Sicily on 4 February 1949; some F4U-4 "Corsair" fighters are in the foreground. Source: www.history.navy.mil

Chapter 9 - CVB's: The Battle Carriers

'The life of the Midway also demonstrates the progress of our Navy; the accommodation of our ships to aircraft of high performance; the use of missiles; exploitation of electronics; the capability to employ a whole family of weapons unheard of when her keel was first laid. No other navy, no other service of any country has a single military unit as powerful, as versatile and as mobile as this great ship.'—V.Adm., George W. Anderson, Jr., Chief of Staff, U.S. Pacific Command, 1957.

Like the CVE's, the CVB's were a direct product of World War II needs and experience, though their missions were different. The former were to be most effective in providing close in support of troop landings. The latter was designed to pit against the enemy the most potent aircraft carrier the world had yet seen.

The CVB's were to provide a solution to the problem of designing a tough rugged ship which would have good aircraft operating features as well as every possible characteristic that would enable it to both give and take punishment.

Chapter 9

The Evolution of Aircraft Carriers

Our early war losses were caused by our failure to adequately control damage sustained. It was obvious that we needed a much sturdier aircraft carrier than we operated in the early years of the war, one with an armored flight deck and improved compartmentation. The resulting design gave us a new breed of ship, battle-cruiser fast, battleship rugged, and with more aircraft operating capacity than anything we had known.

At the same time, aircraft designers were producing larger, heavier types to be operated off sea-going carriers. These higher performance planes, heftier, faster, would place great demands on the flight decks of the proposed CVB's. The planes would require greater room, and these considerations added to the overall weight of the constructed carrier.

On July 9, 1942, Congress authorized their construction. Already, the toll on both U.S. and Japanese carriers had been heavy. In January that year, the *Saratoga* was damaged by submarine torpedo and forced to a yard for repairs. In the Battle of the Coral Sea in May, the light carrier *Shoho* was sunk by U.S. carrier-based planes which, the next day, also damaged the *Shokaku*. In this battle, the *Yorktown* was damaged; the *Lexington*, ravaged by uncontrollable fires, sank. During the decisive Battle of Midway, the Imperial Japanese Navy lost the *Akagi*, the *Kaga*, the *Hiryu*, and the *Soryu*, *Yorktown*, already damaged at Coral Sea, was hit again at Midway and on June 7 was sunk.

Midway was a significant victory for the Allied forces. While proving a turning point in the war, it again conclusively demonstrated the warfare potential and, in fact, superiority of carrier aviation. To commemorate the occasion, the escort carrier CVE-63 was named USS *Midway*, but on September 15, 1944, her name was changed to USS *St. Lo*, relinquishing her name to the first of a new class aircraft carrier then being built, USS *Midway* (CVB-41). This battle carrier was laid down on October 27, 1943. A sister ship, CVB-42, was laid down as USS *Coral Sea* on December 1, 1943, but upon the death of the President, was renamed USS *Franklin D. Roosevelt*. The third large aircraft carrier built, CVB-43, became USS *Coral Sea*.

Contracts for the new carriers were signed August 7, 1942, and by September 18, plans for them were well under way. On that date, the Chief of the Bureau of Ships wrote to the Commander in Chief, U. S. Fleet, to the Vice Chief of Naval Operations and to several Bureau chiefs, discussing the proposed contract design:

"It will be noted that the island is shown offset from the side of the flight deck to the maximum extent permitted by clearance for passage of the Panama Canal," he wrote. "This location of the island has the obvious advantage that a straight fore and aft flight deck runway for airplanes is interfered with to the least possible extent. It gives a flight deck width in way of the island of 107 feet."

Chapter 9

The Evolution of Aircraft Carriers

This was one of the last times the Panama Canal was a limiting factor in the construction of aircraft carriers. The "Canal block" was broken when it was later decided to construct a carrier not to go through it.

Concerning the island structure, BuShips continued: "Extensive wind tunnel model tests of the CV-9 class island with a large number of modifications involving various degrees of streamlining and attempts to reduce smoke nuisance on the flight deck caused by stack gases have been performed. These studies showed clearly that the details of island contour were of negligible importance in effect upon air-flow patterns as compared with the bulk of the ship and of the island itself. In view of these conclusions, attempts to streamline the various essential protuberances on the island and of the island itself were discarded in the case of the CV-9 class and, therefore, have not been incorporated in the present plans."

The island structure was the subject of considerable correspondence in the months and years following. There was an obvious effort by most bureaus to keep the island as small as possible. In this there was general agreement. Comment and discussion became extensive when locations of specific spaces in the island were brought up, as well as uses to which they would be put. Occasionally, proposed requirements threatened to bloat the island structure, but as alternate locations were found, it was possible to keep it to a reasonable size. In October 1942, for instance, the Chief of the Bureau of Aeronautics, RAdm, John S. McCain, noted:

USS Franklin D. Roosevelt (CVB-42) launches a Lockheed P2V "Neptune" bomber with "JATO" assist, during a Task Force 21 cruise, 2 July 1951.
JATO stands for Jet-Fuel Assisted Take Off. Source: www.history.navy.mil

Chapter 9

The Evolution of Aircraft Carriers

"Location in the island of the following space, the functions of which do not necessarily require island space is noted: Pilot balloon room, two squadron lockers, repair I, flight deck crew, flight deck control, flight deck equipment, and one unassigned space. This bureau considers that effort should be continued to reduce island size."

The original proposals called for the installation of two flushdeck type catapults capable of launching VT type aircraft and one double action type in the hangar, capable of launching fully loaded VSB type aircraft. But by October 1942, the General Board considered the complications involved in the installation of a hangar catapult and decided against it. Within the year, the decision was reached to eliminate hangar catapults from *Essex* class carriers, then either under construction or planned.

Hangar fires resulting from combat damage offered particular danger in both Japanese and U. S. aircraft carriers during the early days of the war. In designing the CVB-41 class carriers this danger was considerably lessened by the introduction of four bulkheads in the hangar, dividing it into three spaces connected by sliding doors.

Underwater subdivision of compartments and spaces was given considerable attention, in event of torpedo or mine hit, and was described as "excellent." To provide additional protection, the flight deck was armored with 3½ inches of solid steel, and the deck side belt armor at the waterline tapered from 7½ inches to 3.

In 1943, the wave of war in the Pacific turned against the Japanese as Allied forces made a concerted offensive, capturing Rendova Island in July. The Japanese-held airfield at Munda in New Georgia Island was taken by the Allies, who invaded Bougainvillea in October and landed on the Gilberts in November.

That same year, U. S. shipyards launched and the Navy commissioned 15 CV's and 24 CVE's.

In early 1944, the Marshalls were taken. On the first day of this operation, complete control of the air was obtained and maintained by carrier-based aircraft. The Marianas were invaded in June and Guam recaptured in August. Leyte was occupied in October-November, the opening blows struck by Task Force 38 under VAdm. Marc Mitscher. American shipyards, mass production well organized, launched 7 more CV's, 33 more CVE's.

Chapter 9

The Evolution of Aircraft Carriers

"The men then proceeded to a CASU, where they awaited shakedown of a carrier other than their own. Their own still was building. Most of the *Midway*'s original crew leaders shook down on the USS *Antietam* and the USS *Charger*. On this shakedown, embryo plane handlers stood battle stations, observed the regular crew at work and finally assisted. They were supervised by a training officer from ComAirLant who expedited their progress."

"Following this shakedown, the *Midway*'s nucleus crew returned to a CASU near where the ship was building. Here they were groomed in taxiing, spotting and parking aircraft. The work [was] accomplished on a runway painted to simulate a flight deck. Also, they familiarized themselves with the aircraft they would be using."

Midway conducted her shakedown in the Caribbean, devoting 51 out of 57 days to air and gunnery operations, simulating all types of wartime conditions. Exercises included fueling escort ships at sea, damage control drills and problems, A.A. tracking and firing at towed spars and drones, emergency lube-oil drills for engineers, arming planes, gassing, and use of inert gas.

USS Coral Sea (CVB-43) shortly after she was commissioned in October 1947 - The Coral Sea incorporated many of the features learned in WW II and in in the construction of her processors and therefore could withstand a much harder hit by torpedo and had a reduction in fire danger within her hanger deck thanks to compartmentation and thicker steel she possessed.
Source: www.history.navy.mil

Air operations involved all types of flying and battle exercises, climaxing the tour with a two-day strike against the Caribbean island of Culebra—a well-

Chapter 9
The Evolution of Aircraft Carriers

pummeled three-mile tract of land used by U. S. warships for shakedown training at that time.

USS *Franklin D. Roosevelt* also conducted her shakedown training in the Caribbean, under command of Capt. Apollo Soucek. After post-shakedown alterations in New York, she was shifted to Norfolk, where she became flagship of Adm. Marc Mitscher during the first large-scale training operations since the end of World War II. These maneuvers of the Eighth Fleet took place in the western Atlantic between April 19 and May 27, 1946.

In the following year, during Caribbean maneuvers, Sikorsky H03S helicopters were operated. Noted Naval Aviation News in June 1947:

"It was not the first time a helicopter had operated off a carrier deck. Four (of them) were with the Byrd Antarctic expedition. But the helicopter really proved its worth as a utility and rescue plane off the *FDR*, a showing which may have an effect on fleet operations of the future."

Activity of the *FDR* in the early post-war years was typical of that of her sister ships. After an extended yard period between March 1947 and July 1948, she completed refresher training in the Caribbean before leaving for her second tour in the Mediterranean. At this time, the "Berlin blockade" was formed and the presence of CVB-42 in that area provided a "show of strength." This was her mission for the next five years, as the Berlin blockade was followed by crises in eastern Mediterranean countries and armed aggression in Korea.

In October 1952, the CVB's were re-designated attack aircraft carriers (CVA's). In 1953 the fleet modernization program was authorized. First aircraft carrier to undergo rework was the *FDR*. The ships were equipped with steam catapults, hurricane bows, and the angled-deck design of Project 110.

Chapter 10

The Evolution of Aircraft Carriers

Japan's second aircraft carrier to be named Amagi was of the Unryu class; the first was sunk at Midway. She accommodated 65 planes. The Unryu was a departure from the advance Taiho class. With this class, Japan returned to its original carrier designs roots. Speed above armor and protection was the rule instituted on the Unryu. Source: militaryhistory.norwich.edu

Chapter 10 - The End of the 'Bokubokan' In WW II

Japan is beaten, and carrier supremacy defeated her. Carrier supremacy destroyed her army and navy air forces. Carrier supremacy destroyed her fleet. Carrier supremacy gave us bases adjacent to her home islands. . . . Carrier supremacy demolished the island air bases and eliminated the air force which was using them. Carrier supremacy made the island naval bases untenable for such shipping as escaped our subs. Carrier supremacy permitted us to give close, tactical air support to the troops who stormed the island fortresses.'—V.Adm. Marc A. Mitscher, USN, quoted in Naval Aviation News, October 1945

When Japan struck Pearl Harbor on December 7, 1941, she had the strongest aircraft carrier force in the Pacific. This supremacy lasted until June 1942, when the Battle of Midway was fought and won by the U.S. Thereafter, the bokubokan ("mother-ship for aircraft"), though an effective and dangerous fighter, was an ever weakening force; ships sunk by U.S. planes and submarines were not replaced in sufficient numbers and strength. The study of the Japanese maritime wartime construction is a study of desperation in the face of an inevitable defeat.

At the outbreak of war, Japan had six fine bokubokan, the carriers *Akagi*, *Kaga*, *Soryu*, *Hiryu*, *Shokaku* and *Zuikaku*, in addition to three lighter carriers, the *Zuiho*, *Hosho* and *Ryujo*. The keels were already laid for others and some conversions were being made. At that time, the U.S. had only seven carriers,

Chapter 10

The Evolution of Aircraft Carriers

widely dispersed. At the Battle of Midway, Japan lost *Kaga*, *Akagi*, *Hiryu* and *Soryu*– and never fully recovered from this decisive defeat.

USS Essex (CV-9) planes attack two Japanese aircraft carriers at Kure, 19 March 1945. An SB2C "Helldiver" scout bomber is visible in the upper right. Ship at bottom is either Amagi or Katsuragi. The other carrier is Kaiyo. Source: www.history.navy.mil

Japan's first wartime constructed carrier was the *Taiho* ("Big Lucky Bird"), a 29,300-ton ship authorized under the 1939 estimates. Built at Kawasaki Dockyard, she was laid down in July 1941, launched in April 1943, and delivered in March 1944. She had a cruising range of 10,000 miles at 18 knots, but could reach 33 knots with ease. Kawasaki claims her to have been the most heavily protected flattop in the world at the time of her delivery. And well she might have been; her armor was impressive.

Taiho had 3¾ inches of plating on the flight deck between her two elevators, covering a distance of some 164 yards. The platforms on these elevators were

Chapter 10

The Evolution of Aircraft Carriers

Aso, Kasagi nor *Ikoma* was completed by the end of the war. *Aso* was launched November 1, 1944, *Ikoma* on October 17, and *Kasagi* two days later. They were 60% to 80% complete when work on them was halted because of material shortages. *Aso* was used as a target ship for Kamikaze training attacks and did not survive this abuse. *Ikoma* was moored at Shodo Jima where she sustained bomb damage toward the end of the war. She and *Kasagi* were scrapped. Seven more Unryu class ships were added to the 1942 program, but they never got beyond the paper work.

Imperial Japanese Navy Kamikaze Pilots dressed in traditional adorned flight suits with arm bands. What this book's Editor finds hard to understand is how these young men trained with the Aso as a target without killing themselves. As the text above states – "Aso was used as a target ship for Kamikaze training attacks and did not survive this abuse". How could Kamikaze pilots practice dive-bombing into a ship and live to actually carry out this act on another day in actual warfare? Source: www.comicvine.com

The Japanese wartime carrier construction program, though ambitious, was not at all successful. What few successes they did enjoy were short lived. Since the pressure was on— especially after the Battle of Midway —it was natural that they would turn to quick conversions. In this area, too, the results were discouraging.

The submarine depot ships *Taigei, Tsurugisaki,* and *Takasaki* were the first to be converted. They became the *Ryuho, Shoho* and *Zuiho*.

Ryuho's structure was weak when she entered the yard for conversion. While being strengthened and given carrier characteristics, she was hit by several bombs from one of the B-25 bombers led by Jimmy Doolittle and launched from

Chapter 10
The Evolution of Aircraft Carriers

the USS *Hornet*. This, of course, delayed completion. When conversion was completed, she displaced over 15,000 tons standard. She had a speed of 26.5 knots, was armed with eight five-inchers, and accommodated 31 aircraft. *Ryuho* saw much action, participating in the battles of the Philippine Sea and Leyte Gulf in 1944. In March 1945, she was moored at Kure, bombed by carrier-based U.S. aircraft, and gutted by fires.

Shoho and *Zuiho* both displaced over 13,000 tons standard upon completion of conversion. *Zuiho* was completed in December 1940, while *Shoho* was completed nearly two years later. Both had a speed of 28 knots, were armed with eight five-inchers, and accommodated 30 aircraft.

Shoho's first battle was her last: she was sunk by carrier-based aircraft of the *Lexington* and *Yorktown* on May 7, 1942, during the Battle of the *Coral Sea*. *Zuiho* was not much luckier. Her contributions to the Battle of Midway and the Aleutians campaign were negligible. At the Battle for Leyte Gulf, she was sunk by carrier-based aircraft.

The conversions of the *Ise* and *Hyuga* from battleships proved to be one of the most puzzling experiments undertaken by the Japanese after the Battle of Midway. Their aft turrets were removed and abbreviated flight decks were installed. A large hangar, an elevator, and two catapults were added, permitting the ships to launch all her aircraft in 20 minutes.

The planes scheduled for these ships were sent to Formosa before the ships were completed. The conversions were employed in the Battle of Leyte Gulf. By this time, Japan had run out of aircraft to supply them, and the ships were used solely in their capacity as battleships. They were later sunk, in July 1945, by U.S. carrier-based planes.

Another conversion, that of the *Ibuki* from an improved Mogami class cruiser, also had a rough time of it. She was authorized under the 1941 program, but shortly after her launching in May 1943, work on her was halted for six months while authorities haggled with the possibility of reconverting her into a fast oil tanker—much needed by the Japanese navy. The decision made, work renewed, this time at a furious pace. Four of her eight boilers were pulled out and this space used for the storage of fuel oil. A hangar and two elevators were installed, and a bridge was placed on her starboard side. She was capable of 29 knots and could carry 27 aircraft. But work stopped again, this time when the construction of small submarines took priority in the shipyards. She was never finished; at the end of the war the *Ibuki* was scrapped.

The most ambitious conversion and the most disappointing career was that of the Yamato class battleship *Shinano*. Laid down as a battleship but not completed when hostilities broke, the possibility of converting her to a carrier was entertained. This possibility became a necessity after the Battle of Midway. Survivors of this battle pointed out serious deficiencies in carrier construction and designers at the Naval Technical Bureau listened well. Heavier armored

Chapter 10

The Evolution of Aircraft Carriers

flight decks were needed to protect them from dive-bombing attacks. Fuel and ammunition stowage spaces needed redesign.

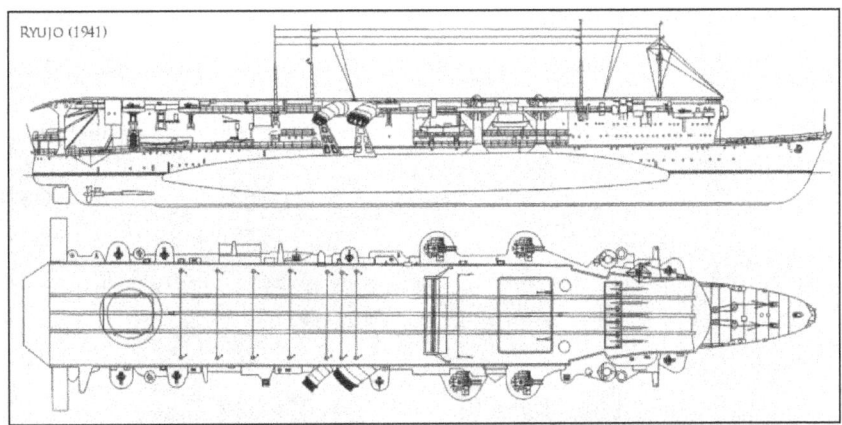

IJN Ryujo was officially classified as an aircraft carrier of the Zuiho class but characteristics differed considerably. During her conversion, she was damaged by Hornet- based B-25 bombers. This sketch shows the single elevator platform in the forward portion of her flight deck and the seven arresting cables on her aft flight deck. Source: www.history.navy.mil

Originally, plans for the conversion of the *Shinano* called for her to act as a "hotel ship," supporting land or other carrier-based planes; she was not to carry aircraft of her own. This plan was changed and by September 1942 the new design was completed and construction began.

Shinano, basically, was to be a CVB. Heavy emphasis was placed on armor. Large bulges below the water line were to minimize torpedo damage. At the water line, an eight-inch thick belt of armor was retained. Four-inch thick armored deck had already been installed before conversion started and this deck became the hangar deck. Rolling metal curtains opened up the forward two-thirds of this deck for night operations and rough seas. The remaining third was closed completely when the curtain was rolled into place. Her flight deck and elevators were designed to withstand 1000-lb. bombs. With this weight, *Shinano* displaced 68,000 tons during her trials at sea.

The Battle of Midway also called attention to the ship's ventilation system. All ducts were protected with 1½-inch armor. Wood was eliminated from the ship wherever possible. A fire-resistant paint was introduced, and a bubble fire-extinguishing system was installed.

The carrier was launched on November 11, 1944 and commissioned November 19th. On the 20th, yard workers still aboard, crewed by green hands, she got underway for Kure where the air complement was to board.

It was at this point that USS *Archerfish* picked her out on radar while surfaced. The submarine maneuvered for position and waited until the carrier and her three-destroyer escort crossed her line of fire. *Archerfish* fired six torpedoes;

Chapter 10

The Evolution of Aircraft Carriers

four hit the carrier. Slowly, she flooded and listed; by 1018 the following morning, all hands were ordered to abandon ship. A few minutes later, *Shinano* capsized and sank—with half her crew still aboard.

For many in the Japanese Navy, the powerful *Shinano* was the last hope. With her sinking, Japanese carrier aviation died, never to operate again.

Japan's most ambitious conversion was the CVB Shinano from the hull of the Yamoto class battleship. Pre-surrender burning of the Imperial Naval Technical Bureau archives destroyed existing photographs of her. This 1950 painting was made by LCdr, Shizuo Fukui - IJN. Source is compliments of: imperialjapanesewarships.devhub.com/blog 499175-cvb-shinano/

Chapter 11

The Evolution of Aircraft Carriers

HMS Ark Royal, a 22,000-tonner, had large hangars on two decks, three elevators. She boasted the largest wardroom in the Royal Navy. In war, her fighters downed or damaged more than 100 enemy aircraft; her bombers wrecked Sardinian airfields, hit Italian Navy, blockaded the French base at Mersel Kabir, Algeria, and hit Italian Navy ships and shore facilities. The ship could carry 60 to 70 aircraft and had 16 x 4.5inch (114 mm) guns. Dimensions were 243.8 x 28.8 x 8.53 meter, speed 31 knots and 1600 crew. Ark Royal sank on 14 November 1941, sinking 30 miles east of Gibraltar after a single torpedo hit from the German submarine U-81, Ark Royal lost all power shortly after the hit and the ship had no auxiliary diesels to drive the fire and salvage pumps.
Source: www.navalofficer.com.au

Chapter 11 - The Wartime European Carriers

'Experience with regard to the suitability of the present type of aircraft carrier must still be evaluated. Examination of enemy naval strategy in ocean warfare leads, however, to the clear recognition of the fact that aircraft carriers or cruisers with flight decks for use in warfare in the Atlantic definitely cannot be dispensed with.'—Gross Admiral Erich Raeder, Commander in Chief Kriegsmarine, during a mid-1940 conference with the Fuehrer on matters dealing with the German Navy.

During World War II, four European nations designed, constructed and/or operated aircraft carriers, or attempted conversions of other type ships to carrier characteristics: Great Britain, France, Germany, and Italy. Great Britain met with extraordinary success, especially in the design of carriers. Among the advances made were the prototype of the WW II-produced CVE (structurally, USS *Langley* qualifies as the first unintended CVE) and experiments that eventually led to the perfection of the "steam slingshot" catapult. Her experiments have a continuing effect on the design of modern carriers. France operated a converted battleship, the *Béarne*, and was building two carriers,

Chapter 11

The Evolution of Aircraft Carriers

Joffre and *Painléve*, when war started. These two carriers were never completed and France fell to the Axis too early in the war for her Navy to make any advancement in carrier aviation. At the same time, Germany's efforts were fitful, frustrated and fated to failure. And Italy, tardily entering carrier-conversion efforts, found the war ended with her ships unfinished.

A starting point in the catalogue of incredible events that launched the nations of the world into global war was the assumption as Chancellor of Germany by Adolph Hitler on January 30, 1933. In the following October he withdrew his country from the disarmament conference and from the League of Nations. Nearly five years later, Germany invaded and annexed Austria. Next on his list was Czechoslovakia in September 1938 which, by skilled "brinkmanship" on the part of the Fuehrer, ended in the Munich agreement. Overconfident now, Hitler zeroed in on Poland. This was too much for both England and France and, on September 3, 1939, they declared war on Germany, and World War II began.

When war began, Britain had six aircraft carriers in commission and six more under construction. Of those operating, the 22,000-ton *Ark Royal* (most recent addition to the Fleet, 1938) and the converted large light cruiser *Courageous* operated with the Home Fleet. The *Furious*, stationed at the Firth of Forth, was used for carrier deck training (but immediately took up convoy duty in the North Atlantic). *Glorious*, sistership to *Courageous* was assigned to the Mediterranean, while the aircraft carrier *Eagle*, which was initially laid down as the dreadnought battleship *Almirante Cochrane* for Chile's Navy in 1913, converted and commissioned an aircraft carrier in 1924, covered the China Station. *Hermes*, the first ship in the world designed from the keel up as an aircraft carrier, also completed in 1924 (the Japanese *Hosho* was completed December 1922), was conducting antisubmarine warfare in home waters.

HMS *Courageous* in 1939 shortly before she was torpedoed by German U-29 with aircraft on approach and sunk with 518 officers and crew out of 1,259 aboard, including her Capt. Source: ww2today.com

Chapter 11

The Evolution of Aircraft Carriers

In addition to the tactical carriers, Britain had one other carrier of lesser, but still significant, capabilities: the *Argus*, worked on between 1916 and 1918 from the Italian liner *Conte Rosso*, was employed on convoy escort duty.

As the political climate changed in Europe and war clouds gathered; Britain made a substantial effort to reinforce her modest and generally venerable carrier fleet. She ordered six new carriers. When the storm broke, these six were in various stages of construction: *Formidable*, *Illustrious*, *Implacable*, *Indefatigable*, *Indomitable*, and *Victorious*. In addition, the 14,500-ton aircraft depot ship, *Unicorn*, under construction in 1939, was to be completed as a CVE.

HMS Formidable – an Illustrious Class Fleet Aircraft Carrier ordered on 19th March 1937 from Harland and Wolff Ltd. at Belfast under the 1937 Program. She was laid down on 17th June that year and launched on 17th August 1939. She was the 6th RN ship to carry the name.
Source: www.naval-history.net

The first years of World War II were expensive ones for Britain's small carrier fleet. *Courageous* was the first carrier casualty of the war. Tracking down a reported U-boat on September 17, 1939, she turned to receive her returning planes when the U-29 submarine plowed torpedoes into her. The carrier sank with nearly half her crew still aboard.

Loss of the *Glorious* was particularly heartbreaking. In June 1940, she participated in the British withdrawal from Norway. Land-based RAF Gladiators and Hurricanes were embarked at Narvik. This was a particularly hairy operation, for none of the planes was configured for carrier landing and the Air Force pilots were not "carqualled" (Carrier – Qualified - Qualified to land on aircraft carriers); all landed safely. Presumed low on fuel, she was ordered to proceed home independently. En route, the carrier was spotted by the German battleships *Gneisenau* and *Scharnhorst* on June 8, and attacked. "Chocked" with RAF aircraft, she was in no condition to launch defending planes. Pounded

Chapter 11

The Evolution of Aircraft Carriers

mercilessly by enemy guns, the ship developed a list and within an hour went down.

The Fairey Firefly was a British carrier-based fighter used during World War II and later. It was involved with a series of missions against the German ship Tirpitz and eventually was used in all theaters of operation during the Second World War.

These losses were balanced in 1940 by the introduction of the *Illustrious* (first of her class) and *Formidable*. They displaced 23,000 tons each, had a length of 753 feet and a beam of 95 feet. They were soon joined by *Victorious*, of the same class, and *Indomitable*, a carrier in a class by herself. The latter had two hangar decks.

An early contribution to carrier operations by *Illustrious* came when she had installed a search radar system for the tracking of enemy aircraft. She was also the first carrier to have a fighter-direction officer aboard. With this effective teaming of men and electronics, *Illustrious*-based planes claimed 75 enemy aircraft in a little over six months of operation.

HMS *Eagle* was the first aircraft carrier to launch planes against enemy surface warships in WW II. On July 9, 1940, carrier-based Swordfish torpedo bombers attacked the Italian fleet in the Med. Defective torpedoes permitted only limited success: only one of the Italian destroyers was sunk. The first successful wartime carrier strike in history occurred on the night of November 11, 1940 when two striking forces from the carrier *Illustrious* attacked the important Italian Naval base at Taranto. Winston Churchill said of this successful raid:

Chapter 11

The Evolution of Aircraft Carriers

"By this single stroke, the balance of naval power in the Mediterranean was decisively altered. The air photographs showed that three battleships, one of them a new Littorio, had been torpedoed, and in addition one cruiser was reported hit and much damage inflicted on the dockyard. Half the Italian Fleet was disabled for at least six months, and the Fleet Air Arm could rejoice at having seized by their gallant exploit one of the rare opportunities presented to them."

HMS Eagle seen here is 1941. On 6 June 1941 her aircraft sank the German merchant ship Elbe (9,179 tons) and on 15 June 1941, in conjunction with the British light cruiser HMS Dunedin, she intercepted the U-boat supply ship Lothringen (10,746 tons) which surrendered about 1000 miles west of the Cape Verde Islands. Soon after this action she returned to England for refitting and was then sent back to the Mediterranean early in 1942. In February 1942 she carried aircraft for Malta and took part in Operations "Spotter" and 'Picket". In June 1942, Sea Hurricanes from 801 Squadron on HMS Eagle provided top cover for Harpoon, the essential supply convoy heading for Malta. Upon entering the Mediterranean, the convoy came under almost constant attack from the Germans and Italians. Source uboat.net/allies/warships

The defeats at Taranto and Cape Matapan (March 30, 1941) finally gave the Italian admirals, who had been pleading for an aircraft carrier since 1925, an effective argument in their dealings with the Italian Air Force which controlled military aircraft. Several plans were actually drawn up but the progress of war did not permit the laying down of keels. Material and manpower shortages forced the Italians to abandon the idea of building carriers from the keel up; instead, they attempted to convert merchant liners.

Early in the war, September 1939, Dr. Joseph Goebbels' Ministry of Propaganda jubilantly reported the sinking of *Ark Royal* by a German bomber. This widely publicized error caused the Third Reich considerable embarrassment, for the

Chapter 11

The Evolution of Aircraft Carriers

carrier escaped undamaged and operated effectively until November 11, 1941, when she finally fell victim to U-boat torpedoes. A month later, HMS *Audacity* met a similar fate. This ship, converted from the German prize *Hannover*, became Britain's first escort carrier upon her completion in June 1941. She was sunk during a battle between U-boats and a Gibraltar-U.K. convoy. Her planes and surface escort destroyed five enemy subs and the decision was made to press for the building of more escort carriers.

Of the losses sustained by the British, *Hermes* was the only aircraft carrier sunk by the Japanese. Fleeing from Trincomalee just ahead of the expected Japanese carrier strike, on April 8, 1942, she was spotted by enemy carrier-based planes. *Hermes*, hit by some 40 bombs, sank in 20 minutes.

Other losses sustained by the Royal Navy included the *Avenger* (November 1942) and the *Dasher* (March 1943), both *Archer* (U.S. Long Island) class escort carriers, *Nabob* was irreparably damaged by torpedo in August 1944 and *Thane* suffered the same fate in January 1945; both were of the Smiter (U.S. *Bogue*) class escorts.

Carrier construction of all types was not pushed in the United Kingdom during WW II in any way comparable to U.S. efforts. Anti-submarine warfare craft had the highest priority and the U.K. depended upon U.S.-built Lend-Lease CVE's (in all, 37) for most of its build-up. Completion of two of the 23,000-ton Implacable class was delayed until 1944. Her sister ship was the *Indefatigable*.

HMS Indefatigable joined the Home Fleet in July 1944 to take part in the attacks on the German Battleship Tirpitz in Norway, with Operation Mascot of 17 July 1944, then subsequently a further series of attacks on the Tirpitz on 22, 24 and 29 August 1944 as part of Operation Goodwood.
Source: www.history.navy.mil

Five carriers of the *Majestic* class and seven of the *Colossus* were laid down, but only the first five of the *Colossus* were completed before VJ day; each displaced

Chapter 11

The Evolution of Aircraft Carriers

14,000 tons. Four of eight of the new 18,300-ton *Hermes* were produced. They were appreciably longer and faster than the *Colossus* class, comparable to the U.S. Navy's first carrier named *Enterprise*. The remaining *Hermes* class was canceled.

Two of the four ships of the new 33,000-ton *Ark Royal* class were laid down but none was completed until well after the end of hostilities.

In addition, the British planned three 45,000-ton *Gibraltar* class carriers (others: New Zealand and Malta), but the project was canceled at the end of the war. These were to be the British equivalent of the U.S. *Midway* class.

During the war, the U.K. operated five light fleet aircraft carriers (the *Colossus* class, in 1945), six fleet carriers of various tonnages, and three escort carriers—all built in British yards—in addition to the ten carriers sunk and the CVE's lend-leased from the U.S. Her carrier-based planes played a vital role in defeating the U-boat offensive. In the Pacific, Adm. Sir Bruce Fraser, RN, commanded the newly established British Pacific Fleet. The 1st Carrier Squadron, comprising the *Indomitable, Victorious, Illustrious* and *Indefatigable*, was a unit of this fleet. Both *Indomitable* and *Victorious* had seen prior action in the Pacific. *Formidable* joined the squadron later. The British group acted as a flying buffer between U.S. amphibious forces and enemy air fields at Sakishima Gunto during the invasion of Okinawa.

Other European powers with carrier aspirations were less successful. France started the war with one converted carrier. The efforts of both Germany and Italy to become carrier powers were foredoomed to failure.

The French carrier *Béarn* was laid down in January 1914 as a battleship of the *Normandie* class. She was finally launched as a battleship in 1920, but three years later entered the yards for conversion to a *Bâtiment Porte-Avions* and was completed in May 1927.

Béarn displaced 25,000 tons, fully loaded, and had an over-all length of 599 feet. She had a complement of 875 and carried 36 to 42 aircraft, including torpedo, reconnaissance and fighter planes. She was held in semi-internment at Martinique from the fall of France in 1940 until 1943. In early 1944 she was taken to the U.S. for rework and emerged as a transport d'aviation, operated by the French.

In 1935, Adolph Hitler announced that his country would construct aircraft carriers to strengthen the Kriegsmarine (the German Navy). The keels of two were laid down in 1936. Two years later, Grand Admiral Raeder presented an ambitious shipbuilding program called the Z Plan, in which four carriers were to be built by 1945. In 1939, he revised the plan, reducing the number to be built to two.

The German Navy has always maintained a policy of not assigning a name to a ship until she is launched. The first German carrier, laid down as Carrier "A",

Chapter 11

The Evolution of Aircraft Carriers

was named *Graf Zeppelin* when launched in 1939. The second carrier bore only the title Carrier "B", since she was never launched. Various names, including *Peter Strasser* and *Deutschland*, were rumored, but no official decision was ever made.

A review of the Fuehrer's conferences on matters dealing with the German Navy, the minutes of which were captured after the fall of the Third Reich, reveals Hitler's vacillating interest in the carriers. Marshall Hermann Goering, Commander in Chief of the Luftwaffe, was resentful of any incursion on his authority as head of the country's air power and he frustrated Raeder at every opportunity. Within his own service, Raeder found opposition in Adm. Karl Doenitz, a submarine man.

Graf Zeppelin, the only one of four aircraft carriers planned by the German navy to be launched, is shown as she appeared in 1939. Never completed, she was captured by the Soviets at the end of the war. Seacocks opened, she rested on the bottom of a shallow channel near Steffin Germany.
Source: listverse.files.wordpress.com

By May 1941, the strain on manpower and raw materials was being felt in Germany. Raeder was still optimistic, however, and informed Hitler that the *Graf Zeppelin*, then about 85 per cent complete, would be completed in about a year and that another year would be required for sea trials and flight training.

Chapter 11

The Evolution of Aircraft Carriers

Though Hitler continued to assure Raeder that the carriers would be built, the Admiral's war with Goering had no truce and became increasingly bitter. Goering showed his contempt for the naval air arm by informing Hitler and Raeder that the aircraft ordered for the *Graf Zeppelin* could not be available until the end of 1944. Goering's tactic was a delaying one—and it worked.

Construction on the carriers had been fitful from the start. Carrier "B" was abandoned in 1940 and broken up. Manpower and material shortages plagued the *Graf Zeppelin*.

Prodded by Raeder, Hitler ordered Goering to produce aircraft for the carrier and under this pressure, the air marshal offered redesigned versions of the Ju-87B Stuka and the Bf-109E-3 fighter which were at that time being phased out of the Luftwaffe first line squadrons. Raeder was unhappy, but he had to accept them or none at all. This forced another delay in the construction of the carrier: the flight deck installations had to be changed.

By 1943, Hitler had become disenchanted with his Navy. Raeder was relieved at his own request and Doenitz, the submarine admiral, took the top naval post. This effectively ended the *Graf Zeppelin* and work on her stopped.

Had the carrier been completed, she would have displaced 23,000 tons, had a length of 920 feet and a beam of 88 feet. Powered by geared turbines, she was to have a speed of 33.8 knots. Her aircraft complement was to have been 42, consisting of ME 109T fighters and JU 87C dive bombers (new designations for the redesigned aircraft). She was to have four screws—unusual for the triple-screw-minded Germany.

The fate of the *Graf Zeppelin* was as stormy as her conception and berth pangs. Scuttled by the Germans, she was raised from the back-water channel near Steffin, by the Soviets in 1946-47. Loaded down with loot, she was towed into the Baltic in 1947, headed for Leningrad. East of Rügen, the ship sank.

Aquila, an attempt by the Italian Navy to convert a liner into an aircraft carrier, but repeated bombings by Allied aircraft never permitted her completion. Many of her parts were cannibalized from the Graf Zeppelin. She is shown as she appeared at La Spezia in 1949, before being scrapped. Source: nonsei2gm.blogspot.com

Chapter 11

The Evolution of Aircraft Carriers

With Germany's abandonment of aircraft carriers came Italy's growing interest in them. The liner *Roma* was earmarked for conversion and many parts of the *Graf Zeppelin* were transported to Italy for use in the conversion. Of particular interest, according to eminent naval historian S.A. Smiley, were the new engines in the ship. Four independent sets of geared turbines from the light cruisers *Cornelio Silla* and *Paolo Emilio* were installed, giving her a designed speed of 30-31 knots. This, says Smiley, was "a unique marine-engineering pearl." The ship's name was changed to *Aquila* and was nearly ready for trials when Italy surrendered. *Aquila* was sabotaged to prevent the Germans from operating her. She was repaired later, but was damaged in two air raids, one in 1944 and the other in 1945. Finally, in 1949, she was towed to La Spezia and scrapped.

Another Italian effort to produce an aircraft carrier by conversion was made when the liner *Augustus*, a running-mate to the *Roma*, was put in hand for conversion in March 1944. She was first named *Falco* and then *Sparviero*, but was never completed. Her half-finished hull was bombed and sunk at Trieste at the close of the war.

A condition of the peace treaty signed in 1947 after a five-week meeting of the Big Four Foreign Ministers in New York specified that no battleship, aircraft carrier, submarine or specialized assault craft could be constructed, acquired, employed or experimented with by Italy, blocking her efforts to be an aircraft carrier nation.

French aircraft carrier Bearn was the only carrier France had completed before the start of WW II. Converted to an aircraft carrier between 1923 and 1927, she had a speed of 21.5 knots or a range of 6000 miles at 10 knots. She spent most of the war years at Martinique in the Caribbean. Source: www.armed-guard.com

Chapter 12

The Evolution of Aircraft Carriers

USS Coral Sea was the last WW II-built carriers to be reworked extensively in the modernization programs; shown here after Project 110A. Source: www.history.navy.mil

Chapter 12 - The Turbulent Post-War Years

'There has been a spectacular advance in aircraft design technology. The transition from propeller-driven aircraft to jet power has been fast. We are now undergoing another evolution from subsonic to supersonic speeds at higher altitudes. By modernization we have utilized our assets of World War II Essex class carriers to the maximum. This has been a military necessity in order to maintain an acceptable degree of combat readiness economically in about half the time required for new construction. Carrier modernization has been pushed vigorously.'—Adm. Arleigh Burke, U.S. Navy, CNO, 1957.

The post-war era was one of dynamic change. The aircraft carriers reflected that change with many modifications designed to equip them to operate the most modern aircraft capable of delivering nuclear weapons and launching guided missiles.

Technological developments were making the *Essex* class obsolescent. On June 4, 1947, the Chief of Naval Operations approved new aircraft carrier characteristics to be incorporated in an improvement program titled Project 27A. This was the first of a series of modernization efforts to modify the *Essex* carriers to meet changing operating requirements.

USS *Oriskany* (CV-34) was the first of the *Essex* class carriers modernized under Project 27A. She entered New York Naval Shipyard in October 1947. At

Chapter 12

The Evolution of Aircraft Carriers

spaced intervals, she was followed by *Essex* (CV-9), *Wasp* (CV-18), *Kearsarge* (CV-33), *Lake Champlain* (CV-39), *Bennington* (CV-20), *Yorktown* (CV-10), *Randolph* (CV-15), and *Hornet* (CV-12). These programs were conducted at Puget Sound and Newport News, in addition to the New York Navy Yard. The *Hornet*, last to be modernized under 27A, left the New York yard in October 1953.

The principal changes involved in the 27A project were directed toward a capability of operating aircraft of up to 40,000 pounds gross weight. The H4-1 catapults were removed and H-8's installed, permitting the launching of considerably heavier aircraft than the carrier had been capable of during the war years. The flight decks were strengthened and the five-inch guns on the flight deck were removed to decrease topside weight, to provide more deck space for parking planes, and to increase safety aspects of the landing area. A special weapon capability was given the last six of the nine carriers modernized under this project. Elevator capacities and dimensions were increased to accommodate heavier planes. And special provisions for jet aircraft were installed—such as jet blast deflectors, increased fuel capacity, as well as some modern jet fuel mixers.

Three of the ready rooms for pilots in these carriers were moved down below the hangar deck, relocating them from spaces directly under the flight deck. This increased pilot comfort and provided better protection. To get the equipment-laden pilots up to the flight deck, an escalator was installed abreast of the island. This provided a single route for pilots manning their planes; it prevented confusion from ship's company rushing up the normal access routes to man battle stations.

In April 1947, *Franklin D. Roosevelt* entered the yards on Ship Improvement Program No. 1, which provided her with a special weapon capability. Her sister ships, the battle carriers *Midway* and *Coral Sea*, followed. This program was also extended to the *Oriskany*, *Essex* and *Wasp*, which had not received the capability under 27A.

Almost a year before the *FDR* entered the yards, the first U.S. testing of the adaptability of jets to shipboard operations were conducted aboard, on July 21, 1946. Successful landings and takeoffs in an FD-1 Phantom were made by LCdr. James J. Davidson.

The Navy continued to experiment with heavier aircraft launchings from carrier decks. In March 1948, carrier suitability of the FJ-1 Fury jet fighters was tested on board the *Boxer* (CV-21) off San Diego. A number of takeoffs and landings were made by Cdr. Evan Aurand and LCdr. R.M. Elder of Fighter Squadron 5A. The following month, Cdr. T.D. Davis and LCdr. J.P. Wheatley made JATO takeoffs in P2V Neptunes from the deck of the *Coral Sea* off Norfolk. This was the first carrier launching of planes of this size and weight.

Chapter 12

The Evolution of Aircraft Carriers

USS Boxer (CV-21) Slides down the building ways, during launching ceremonies at Newport News Shipbuilding & Dry Dock Company, Newport News, Virginia 14 December 1944. Note the banner spread across the front of Boxer's flight deck, proclaiming: "Here We Go to Tokyo! Newport News Shipyard Workers' War Bonds Help to Sink the Rising Sun". Another interesting aspect of this photograph is that it displays virtually all men and woman wearing hats which was the style of the day and everyone is dressed up for the occasion. Courtesy of William H. Davis
Source: U.S. Naval Historical Center Photograph.

Chapter 12

The Evolution of Aircraft Carriers

USS Boxer (CV-21) sailing in Korean waters. Source: U.S. Naval Historical Center Photograph

It was inevitable, then, that the Navy would introduce all-jet squadrons to carrier operations. On May 5, 1948, Fighter Squadron 17-A, equipped with 16 FH-1 Phantoms, became the first carrier-qualified jet squadron in the U.S. Navy. It took three days of operations to do it, but all squadron pilots, in addition to Commander Air Group 17, qualified on the USS *Saipan* (CVL-48), with a minimum of eight landings and takeoffs each.

Project 27A was originally intended for more than nine carriers, but development of the steam catapult and the prospective employment of more advanced types of aircraft made it apparent that this project had to be modified to meet future needs. Accordingly, Project 27C was initiated.

Hancock CV-19) was the first carrier to receive the C-11 steam catapult. This photograph was taken Off San Diego, California, 11 February 1975, shortly before beginning her final deployment to the western Pacific. Source: www.history.navy.mil

Chapter 12

The Evolution of Aircraft Carriers

Hancock, *Intrepid* and *Ticonderoga* were slated for this program—later identified as Project 27C (axial deck). Most important of the changes was the introduction of the steam catapult developed by the British. In 1952, tests of the catapult installed in the Royal Navy carrier HMS *Perseus* were conducted at the Naval Shipyard, Philadelphia, at NOB Norfolk, and at sea during the first quarter of the year. This report was given at the time by Naval Aviation News:

"The new catapult fared so well during the tests that the Navy has already begun an investigation into the adaptability of it to their new flush deck carrier USS *Forrestal*, which is now under construction."

"The new catapult, invented by a Royal Navy volunteer reserve officer, Cdr. C.C. Mitchell, O.B.E., of Messrs. Brown Brothers & Co., Ltd., Edinburgh, uses the principle of the slotted cylinder, and has no rams or purchase cables. A hook on the aircraft to be launched is connected directly to a piston which is driven along the cylinder by high pressure steam from the ship's boilers. A novel sealing device is used to keep the slotted cylinder steam tight."

"While the amount of steam required for sustained operation is large, tests have shown that the boilers can meet the demand without interfering with the ships' operations."

The *Hancock* was the first U.S. carrier to receive the new "steam slingshot," designated C-11 by the U.S. Navy. On June 1, 1954, Cdr. H. J. Jackson, in an S2F-1, was catapulted from the *Hancock* in the initial U.S. operational tests. Throughout the month, testing continued. A total of 254 launchings were made with the S2F Tracker, AD-5 Skyraider, F2H-3 Banshee (Banjo), F2H-4 Banshee (Big Banjo), FJ-2 Fury, F7U-3 Cutlass, and F3D-2 Skyknight aircraft.

In addition to the C-11 Steamcat, Project 27C (axial deck) also provided for a strengthening of the flight deck. The number three centerline elevator was replaced with a deck edge type of greater capacity. Other improvements were made, in addition to those proved efficient in 27A.

Even as these changes were being built in the Hancock, *Intrepid* and *Ticonderoga*, the Bureau of Aeronautics proposed, in mid-June 1952, that a new design flight deck be installed in the *Antietam*. The previous May, both jet and propeller type aircraft were tested on a simulated angled deck aboard the USS *Midway*. The idea was originated by the British and proved very effective for them. *Antietam*'s deck was to extend outboard on the port side from the normal flight deck, thus allowing aircraft landings to be angled 10° off the ship's centerline.

Chapter 12

The Evolution of Aircraft Carriers

Antietam tests British-designed angled deck in the Virginia Capes area in April 1953. Fifteen types of aircraft were used during the evaluation period. Pilots were enthusiastic, for it eliminated barriers, barricades, and danger of parked planes at runway's end. Source: www.navy.mil

Pushed through the guidance design stage by the Hull Design Branch of BUSHIPS in early July, *Antietam*'s new deck was completed in mid-December at the New York Naval Shipyard. At first called a canted deck; this term officially gave way to the more familiar angled deck by OPN AV Notice 9020 on February 24, 1955. It also outlawed the use of "slanted" and "slewed" in describing the deck design.

In December 1953, BUSHIPS Journal reported:

"The final detailed report on the evaluation of the canted flight deck installed in USS *Antietam* (CVS-36) reveals that the operational trials have met with a high degree of success. The canted deck aircraft carrier appears to provide the safest, most desirable, and most suitable platform for all types of aircraft—those currently in use as well as those still on the design board—and is superior to the axial flight deck carrier in these respects."

"The canted flight deck on *Antietam* was finally installed at an angle of 10.5° to the centerline of the axial flight deck. The landing area of the canted deck is 525 feet long with a width at the landing ramp of 70 feet and narrowing to 32 feet, 8 inches, at the extreme forward end of the takeoff area. This gives the effect of 'flying into a funnel,' causing the pilot to head toward the canted centerline. This effect aids him in maintaining the flight and deck path which fully utilizes the complete length of the canted flight-deck."

"Fifteen types of aircraft including both propeller and jet-propelled participated in the tests which were conducted in four phases; extending

Chapter 12

The Evolution of Aircraft Carriers

from December 29, 1952 to July 1, 1953. A total of 4,107 landings were made, including touch-and-go and arrested landings, during day and night operations. During the entire evaluation period there was no major accident and only a total of eight minor accidents, none of which could be attributed to the canted deck principle."

The advantages were immediately manifested. By eliminating the centerline elevators and using one or more deck edge elevators (not installed in the *Antietam*), more elevators would be available for bringing up spares from the hangar and striking "dud" aircraft below. Once landed, the plane could easily taxi onto a starboard deck edge elevator without impeding flight operations.

It was also possible to catapult aircraft and land them simultaneously, and to launch CAP and interceptors on short notice. This gave the carrier improved combat readiness.

The pilots were impressed. An extra margin of safety was given them by removing the danger of crashing into gassed and armed planes parked forward of the landing area. The BUSHIPS Journal commented:

"The clear deck ahead on every carrier pass relieved the pressure on the pilot. Primarily for this reason, pilots who have flown from the canted deck are unanimous in their favorable enthusiasm. This was found to be especially true when *Antietam*'s canted deck was rigged to simulate a CVE type carrier. Pilots flying AF type aircraft confirmed that part of the mental strain of carrier landings is relieved with removal of the barriers, making the landings considerably less difficult."

"Fewer cross deck arresting pendants and arresting gear engines are required for the canted deck. It is considered desirable to keep the landing area as far aft as is practical and safe, yet far enough forward to decrease rates of descent. This can be accomplished only by limiting the pendants to a minimum commensurate with safety and picking optimum pendant locations. Fewer pendants also result in a decrease in topside weight."

Project 27C (angled deck), which resulted from the *Antietam* tests and modified the original 27A, significantly changed the silhouette of the aircraft carriers. The canted or angled deck was installed and the hurricane bow of the original *Saratoga* and *Lexington* carriers reintroduced. The project also allowed for the improvement of the Mark 7 arresting gear by reducing the number of deck pendants by one-half and thereby cutting the ratio of arresting gear sheaves to two to one. The forward centerline elevator was enlarged. Air conditioning and sound proofing made the island spaces more comfortable and efficient. The latest advancements in deck lighting were also installed in these attack carriers.

Lexington, Shangri La, and *Bon Homme Richard* all received the improvements of this project and they were so successful that *Hancock, Intrepid* and *Ticonderoga* returned to the yards for this new conversion.

Chapter 12

The Evolution of Aircraft Carriers

The trend extended, inevitably, to the *Midway* class. In September 1953, the Navy announced new modernization plans for these carriers under a new program called Project 110. In May 1954, the *Franklin D. Roosevelt* entered Puget Sound Naval Shipyard for the conversion. *Midway* followed in September 1955. These carriers received the best features of the 27C (angled deck) conversion which were incorporated in Project 110. Additionally, they had a modified steam catapult installed in the angled deck area; full blisters were added for maximum protection, liquid stowage, and stability, and the after starboard elevator was relocated to the starboard deck edge.

With the changes in carrier configuration ran a parallel change in missions and these changes were reflected in the re-designation of certain carriers as they appeared in the Navy Vessels Register.

USS Oriskany (CVA-34) became the first of the carriers to be reworked in the post-war modernization program. Angled deck was installed in Project 125A. This photo was taken off the San Francisco Naval Shipyard, California, on 27 April 1959, following installation of her new angled flight deck and hurricane bow. Source: www.history.navy.mil

On October 1, 1952, the very familiar CV and CVB designations went by the board. The ships were assigned the designation CVA, reflecting their reclassification as attack carriers. Prior to this, only the CV's were known as attack carriers, in the Fleet, to distinguish them from the CVB's. Antisubmarine Support Aircraft Carriers became a new classification in July 1953 and was applied to those attack carriers assigned to ASW; the following August 8, five CVA's were re-designated CVS's, ASW support carriers.

There were no further changes in designations over the next two years, but in July 1955, *Thetis Bay* (CV-E 90) became CVHA-1. This proved the first move in the eventual disappearance of escort carriers from the operational Fleet. The attempt to modify CVE's for a new role in helicopter vertical assault operations

Chapter 12

The Evolution of Aircraft Carriers

was abandoned when the experiment proved too costly. On May 7, 1959, the classification of 36 escort carriers, designated CVE, CVU, and CVHE, was changed to AKV, for Cargo Ship and Aircraft Ferry. New hull numbers were assigned. This ended the role of escort carriers as combat ships of the Fleet.

On December 30, 1957, USS *Saipan* (CVL-48), last of the light carriers, was decommissioned. On May 15, 1959, that designation was stricken from the register when the classification of four support carriers, CVS's, and seven light carriers, CVL's, was changed to Auxiliary Aircraft Transport, AVT.

The modernization of individual carriers reflected Navy thinking, Navy accomplishment, and navy planning. The programs were successive steps in what somebody once called "a schedule of orderly retirement." As the carriers aged (some aged "faster" because of battle damage in WW II), they were transferred from the CVA designation to the CVS, then to LPH and retirement, and it all was tied to new construction programs which made it possible to keep the number of operating CVA's up to the prescribed limits. As each new ship was acquired, it took the top position among the CVA's while the one in the bottom position moved to the top of the next lower class.

USS Thetis Bay (CVE-90) was re-designated CVHA-1 and started the phasing out of CVE's from the Navy's list of combatant ships. Here the ship is being converted for the operations of helicopters. She was recommisioned on 20 July 1956, Capt. Thomas W. South, II, in command, and completed conversion six weeks later on 1 September. Source: www.history.navy.mil

The new CVHA-1 carrier arrived at her new home port, Long Beach, on 20 September 1956. There, helicopter teams from Marine Corps Test Unit No. 1,

Chapter 12
The Evolution of Aircraft Carriers

Camp Pendleton, demonstrated landing and take-off techniques. Thetis Bay participated in amphibious training exercises off the California coast before deploying to the Far East on 10 July 1957. She returned to Long Beach on 11 December 1957 and resumed local operations. On 28 May 1959, her designation was changed to LPH-6, amphibious assault ship.

USS *Coral Sea* (CVA-43) was the last aircraft carrier of World War II design to be extensively reworked during the post-war modernization program. She entered the Puget Sound shipyard on April 15, 1957, and was re-commissioned January 25, 1960. In the interim, changes made in her configuration were contained in Project 110A, a modification of the 110 of her sister ships, *FDR* and *Midway*.

The basic changes were the same as those in Project 110, but 110A added new features. Of the three deck edge elevators installed, for instance, one was placed on the port side near the LSO platform. This eliminated the hazardous arrangement of having an elevator contiguous to the landing area. It also simplified maintenance problems and provided the capability of operating all three elevators during flight operations.

Existing arresting gear was replaced with five Mk 7-2 pendant and barricade engines with the new sheave and anchor dampers. *Coral Sea* was the first to have installed, in the fantail area, a complete jet engine test facility; they are now installed in all new carriers. She had twice as much stowage for JP-5 fuel as her sister ships, over a million gallons, in addition to a 62,000 gallon capacity for avgas. And although Ranger was the first to have fuel centrifugal purifiers installed, she did not rely on them exclusively. When *Coral Sea* deployed to WestPac, she had four of them installed and did use them exclusively. During the first 8½ months of operation, she burned approximately seven million gallons of JP-5, according to Air Officer Cdr. D.W. Houck, and did not experience one case of contaminated jet fuel.

Modular CIC, a clock-like layout of communications, radar, and other CIC elements, had been tested in the *Oriskany* and proved successful. It was installed in *Coral Sea*, which became the second aircraft carrier to have such an arrangement.

The modernization program extended the lifetime usefulness of the *Essex* class carriers built during WW II and permitted them and other class carriers to operate jet-powered aircraft of increasing designed power without compromising combat readiness of the Fleet. The important limiting characteristics of the planes operating from carriers are landing speed, landing weight and required end speed, and—in wooden deck ships—the wheel loading.

Chapter 12

The Evolution of Aircraft Carriers

Many new developments have had a profound effect on carrier aviation. In August 1955, for instance, the constant run-out method of controlling arrestment was used in the Mk.5 arresting gear installed in USS *Bennington*. Its primary advantage was the ability to arrest a plane with a minimum amount of hook loads. With the earlier pressure types of controls it was necessary to stop the aircraft in shorter run-out in order to take care of inadvertent over-speed of the aircraft. This put a considerable strain on the planes. The new system is set for the weight of the landing aircraft, so that a 60,000 pound plane would pull out no more wire than a 10,000-pounder.

Mirror Landing System was first tested on the U.S. carrier Bennington in 1955. Photograph above is of USS Bennington as she passes the wreck of USS Arizona (BB-39) in Pearl Harbor, Hawaii, on Memorial Day, 31 May 1958. Bennington's crew is in formation on the flight deck, spelling out a tribute to the Arizona's crewmen who were lost in the 7 December 1941 Japanese attack on Pearl Harbor. Note the outline of Arizona's hull and the flow of oil from her fuel tanks.
Source: www.history.navy.mil

Other pilot aids include TACAN (Tactical Air Navigation System) which gives pilots bearing and distance from a carrier, the British-developed mirror landing system (improved by the use of Fresnel lenses), and PLAT (Pilot/LSO Landing Aid Television).

"We are limited by how far we can go in modernization programs by the age of the ship," said Adm. Arleigh Burke in 1957. "They are getting old.

Chapter 12

The Evolution of Aircraft Carriers

Their machinery is wearing out and they are becoming progressively more expensive to maintain. Like an old car, they must be replaced."

"The modernization programs have been the proving ground for the advances which have been made in carrier operating techniques. But the full combat effectiveness of these developments can be realized only in new construction."

Two years earlier, in 1955, USS *Forrestal* (CVA-59) was commissioned, the first of a new class aircraft carrier. It was a logical step in the evolution of one of the Navy's proven and powerful aircraft weapons systems—the modern ship-of-the-line in the Fleet.

USS Forrestal (CVA-59) Awaits her turn to refuel, while operating in the Mediterranean Sea during the Jordanian crisis, 29 April 1957. USS Caloosahatchee (AO-98) is ahead, with USS Lake Champlain (CVA-39) and USS Salem (CA-139) alongside. Note Forrestal's eclectic air group, with F3H-2N, FJ-3M, F9F-8B, F2H-2P, A3D-1, AD-6, and S2F aircraft visible on her flight deck. Source: www.history.navy.mil

Chapter 13

The Evolution of Aircraft Carriers

USS Enterprise (CV AN-65), the most powerful nuclear-powered aircraft carrier is the first and only such carrier in any Navy of the world. Photographed here many years later on one of her dozens of deployments throughout the world; the world's first nuclear powered aircraft carrier is still prowling the oceans and seas of the world to keep us safe from terrorism. Source: www.navy.mil

Chapter 13 - CVA's Built to Meet Modern Needs

'Events of October 1962 indicated, as they had all through history, that control of the sea means security. Control of the seas can mean peace. Control of the seas can mean victory. The United States must control the seas if it is to protect our security and support those countries which, thousands of miles away, look to you on this ship and the sister ships of the United States Navy.' - President John F. Kennedy, addressing the crew of USS *Kitty Hawk* (CVA-63) on June 6, 1963

The dramatic events of October 1962 to which President Kennedy referred were the missile build-up in Cuba and the immediate U.S. reaction to this threat. This was one of a series of incidents occurring since World War II that endangered the democratic way of life; incidents effectively neutralized by the presence of powerful U.S. carrier forces in the area.

The versatility of the current U.S. carrier fleet is largely due to the operation of what the press has labeled "super-carriers," heavy duty aircraft carriers of the size, power, and potency of the *Forrestal* and the nuclear-powered *Enterprise*. They had a difficult birth.

Chapter 13

The Evolution of Aircraft Carriers

In April 1945, owing to lessons learned from their experience in combat, Carrier Task Force Commanders requested heavier and larger aircraft to accomplish war missions. An informal board was appointed to consider the carrier requirements of the U.S. Navy. The hulking CVB's of the *Midway* class, which were readying for commission and combat duty, provided a stopgap supply to the needs of the Task Force commanders. The Ship Characteristics Board made various studies of the problem, and it was decided that the project should be made a design study for the 1948 shipbuilding and conversion program. Given the designation "6A Carrier Project," one of the carriers was slated to be built in the 1949 construction program.

Between 1945 and November 1948, some 78 different designs were made before final acceptance. On June 24, 1948 Congress passed the Naval Appropriations Act of 1949. This provided funds for construction of the carrier. The contract was awarded Newport News Shipbuilding and Dry Dock Company.

In the planning stage, the new carrier was to weigh 65,000 tons and have a 1030-foot flight deck, a 130-foot waterline beam, and four catapults. Architects went back to original *Langley*, *Ranger* and *Long Island* designs by sweeping the flight deck clear of an island structure. Instead, the carrier was to have had a small island on an elevator apparatus, to be lowered during flight operations. This was one answer to a BuShips objection to the flush deck design, predicated on the fact that a satisfactory method of disposing of stack gases had not been developed.

All elevators were to be along the sides of the ship, with a large elevator at the extreme after end of the flight deck. Added strength of the flight deck was to be made possible by reducing the openings in the hangar sides, so that the ship, from the keel to the flight deck, could be considered as a unit, from the standpoint of strength. This would permit the operation of aircraft well over 100,000 pounds. Adm. Marc Mitscher greatly influenced formation of the project, having been one of the Task Force commanders who recommended heavier, more versatile carrier aircraft.

In July 1948, construction of the carrier was approved by Congress and President Truman. In March the following year, the President authorized the name for the new carrier; when commissioned, she would become USS *United States* (CVB-58).

Chapter 13

The Evolution of Aircraft Carriers

The events of April 1949 occurred with stunning swiftness and to this day are subject of discussion in some military and political circles. On April 13, funds were approved by the House of Representatives. Two days later, Secretary of Defense Louis Johnson wrote to General Eisenhower, then temporary presiding officer of the Joint Chiefs of Staff, requesting that the Joint Chiefs review the need for a new aircraft carrier. At that time, criticism of the entire concept of carrier warfare was again voiced by some military leaders. The carrier's keel

The first of her class, Forrestal profited from lessons learned from post-war designs, particularly from the cancelled CVB-58. Source: www.history.navy.mil

was laid at Newport News on April 18. On April 23, the views of the Joint Chiefs were sent to SecDef and on that same morning Secretary Johnson ordered work on the carrier stopped. Secretary of the Navy John L. Sullivan resigned in protest the next day.

There was no new carrier construction in 1950. However, mid-year events caused Navy planners again to renew requests for heavy-duty carriers. On June 25, 1950, North Korean forces invaded the Republic of Korea. Two days later, President Truman announced he had ordered sea and air forces in the Far East to give support and cover to Republic of Korea forces and ordered the Seventh Fleet to take steps to prevent an invasion of Formosa. On July 3, carrier aircraft went into action in Korea. USS *Valley Forge*, with Air Group Five, and HMS *Triumph*, operating in the Yellow Sea, launched strikes on airfields, supply lines and transportation facilities around Pyongyang, northeast of Seoul.

On July 12, 1951, the Navy Department announced a contract for a new large aircraft carrier (CVB-59), to be built at Newport News Shipbuilding and Dry

Chapter 13

The Evolution of Aircraft Carriers

Dock Company. On July 30, Congressional action approved the contract. A joint resolution from Capitol Hill proclaimed:

"Be it resolved that when and if the United States completes construction of the new aircraft carrier known as the *United States*, the construction of which was discontinued April 23, 1949, or the aircraft carrier authorized in Public Law 3, Eighty-Second Congress, first session, it shall be named the *Forrestal*."

At Newport News, the new carrier was designated Hull Number 506. Her keel was laid on July 14, 1952.

Mr. Charles P. Roane, Supervising Naval Architect, Aircraft Carrier Type Branch, BuShips, commented on the *Forrestal* in the November 1952 issue of BuShips Journal:

"The *Forrestal* incorporates all of the developments from the other carriers, plus those learned from the United States. The increase in size of the *Forrestal* over the *Midway* class comes about as a normal development in aircraft carrier design. With four catapults instead of the usual two and four airplane elevators instead of the usual three, aircraft operations from this ship will be greatly improved."

"The new design was planned to meet added requirements, such as the servicing and starting of jet aircraft, maintaining the electronic appliances on the aircraft in a ready-to-go condition while the plane is on the deck, blending of aircraft fuels to get a fuel which can be used in jets without sacrificing the gasoline capacity, and a flush deck where the navigating bridge can be lowered or raised to suit operating conditions. Stacks comparable to the *Ranger* will be used. New type steels, the result of years of development, will go into the construction."

The flush deck design barely left the drawing board before it was changed. This design was advanced to provide optimum landing area and to eliminate the hazard of island superstructure offered by the axial deck. At the end of W.W. II, however, the British developed the angled deck concept and operated lightly constructed twin-engine attack planes from the marked-off deck of a British carrier. U.S. Navy pilots conducted similar test on the *Franklin D. Roosevelt* and the decision to modify the flight deck of a U.S. carrier was made. Accordingly, the *Antietam* was reconfigured, landings and takeoffs were made using a variety of aircraft, and a final detailed report on the evaluation of the "canted" or angled deck revealed that the operational trials met with a high degree of success. As a result of these experiments, the Navy ordered a redesign of the deck and operating arrangements on the *Forrestal* and all future carriers, as well as reconfiguring many of the existing carriers during scheduled modernization periods.

When Secretary of the Navy Dan A. Kimball announced the awarding of a contract to Brooklyn Naval Shipyard for the construction of USS *Saratoga* (CVA-60), he said it would be similar to the *Forrestal*. But design

Chapter 13

The Evolution of Aircraft Carriers

improvements in machinery since *Forrestal* installation were ordered to give *Saratoga* a somewhat higher speed.

"The importance attached to this carrier [Saratoga] by the Navy Department," Secretary Kimball said, "is emphasized by the Navy's sacrifice of other combatant ships in the 1953 program in order that a second large carrier can be added to the Fleet."

"Although the ships sacrificed are urgently needed to augment the battle readiness of the Fleet, the Navy decided that the need for the large aircraft carrier is even more urgent in terms of national security."

USS Saratoga (CVA-60), sister ship to (CVA-59), was designed to have greater horsepower than Forrestal. She was also designed with four high powered steam catapults and was the premier aircraft carrier of her day. Source: www.history.navy.mil

Forrestal was launched on December 11, 1954, and christened by Mrs. James Forrestal. The ship was commissioned at Norfolk Shipyard on October 1, 1955. The carrier had an overall length of 1036 feet, a width of 252 feet, and nearly four acres of flight deck. She displaced 59,650 tons and had a horsepower rated over 200,000 and a speed over 30 knots. Four steam catapults were installed. She had a complement of 3500 officers and men, including the air group.

Assistant Secretary of the Navy (Air) James H. Smith, Jr., spoke at the commissioning ceremonies. "If our way of life is to survive," he said, "we must maintain these two alternate military postures: the first is to maintain a powerful and relatively invulnerable reprisal force which will signal a potential enemy to stop, look and listen before he risks an all-out atomic war. The second is to insure that we ourselves will not be forced to change

Chapter 13

The Evolution of Aircraft Carriers

the character of a limited war because of fear of ultimate defeat in a series of them. Fortunately, we need not maintain a completely separate set of forces for each posture. In this ship and the variety of aircraft she can service we combine the two, and we add the multiplier of the ability to appear quickly at any one of the many far-flung trouble spots. This is economy of force, achieved without sacrifice of our objectives."

USS Ranger's after deck was altered slightly to provide a longer overall length. Source: www.history.navy.mil

USS Independence was commissioned at New York Naval Shipyard, the fourth aircraft carrier of the Forrestal class. This aircraft carrier had increased arresting gear capability installed which enabled her to launch and recover larger and heavier jet aircraft. Source: www.history.navy.mil

Chapter 13

The Evolution of Aircraft Carriers

USS *Saratoga* was christened at New York Naval Shipyard on October 8, 1955. A few token feet of water were splashed into the new ship's dry-dock to "launch" her officially. She was essentially similar to *Forrestal* but was designed to develop considerably more horsepower. She was commissioned April 14, 1956.

Sister ship *Ranger* (CVA-61) had one outstanding exception to distinguish her when she was commissioned August 10, 1957. The angle of the after flight deck was altered slightly, giving her an overall length of 1046 feet, as compared to the 1039 of *Forrestal*. Another innovation, an all-welded aluminum elevator, was installed on the port side, replacing the conventional steel types used on other *Forrestal* class carriers. To expedite her building, work was started in a smaller dock. About four months later, when the *Forrestal* was launched, the partially completed *Ranger* hull was floated into the larger facility.

CVA-62, the USS *Independence*, was constructed in Dry-dock 6 at New York Naval Shipyard, her stem at the head of the dry-dock to facilitate material delivery over a truck ramp leading from the head of the dock to the hangar deck at the stern. The island and associated sponson "were not installed in order to avoid blocking off the large traveling crane. In August, the extraordinarily complex job of transferring her to Dry-dock 5 was accomplished smoothly and efficiently.

USS Kitty Hawk (CV 63) receives fuel during replenishment at sea from the Royal Australian Navy auxiliary oiler replenishment vessel HMAS Success (AOR 304) as U.S. Navy guided missile cruiser USS Cowpens (CG 63) steams alongside and guided missile destroyer USS John Paul Jones (DDG 53) trails behind. Source: www.defense.gov

Independence was commissioned at the New York Naval Shipyard on January 10, 1959, the fourth carrier of the *Forrestal* class to join the Fleet.

Kitty Hawk (CVA-63) and *Constellation* (CVA-64) were essentially designed along the *Forrestal* lines but developed into a separate class, the *Kitty Hawk* class. The major difference is missile capability. Both CVA63 and -64 are armed

Chapter 13
The Evolution of Aircraft Carriers

with Terriers. The fuel capacity in the *Kitty Hawks* is a little greater than the *Forrestal's*, while avgas capacity is a little less. The angled part of the flight deck is some 40 feet longer and the catapults and elevators have greater capacities. USS *America* (CVA-66), now being built at Newport News, will have an even longer angled deck than any of the predecessors. Placed in the *Kitty Hawk* class, she is scheduled to be completed in late 1964.

On February 4, 1958, Secretary of the Navy William B. Franke announced that the world's first nuclear-powered aircraft carrier was to be named USS *Enterprise* to perpetuate the WW II carrier and her six predecessors. On that same day, the keel of the carrier was laid at Newport News.

On September 24, 1960, Adm. Arleigh Burke, then CNO, delivered an address during launching ceremonies, in which he described the new carrier.

"This new Enterprise, the largest ship ever built, of any kind, by any nation, will be the eighth Navy ship to proudly bear that name. Her forbearers have left an enviable record; a record of courageous, distinguished service."

"We are looking at a major advance in the art of nuclear engineering. The problems which were solved, the know-how that was developed in order to build this ship, represent a tremendous contribution to our knowledge of the military and industrial uses of nuclear energy."

"Her eight powerful nuclear reactors would enable the Enterprise to cruise 20 times around the world without refueling. Her great endurance and her advanced hull design would allow the ship to make this extraordinary journey at sustained high speed, exploiting to its utmost the seagoing advantage of mobility."

From the very first, it was obvious that designers and builders of Newport's hull No. 546, the *Enterprise*, had hit the jackpot. For the first time, RAdm, F.S. Schultz, Assistant Chief BuShips, noted the customary builder's trials of a major combat ship were eliminated, and the ship was presented to the Navy for acceptance trials on her first trip to sea.

Enterprise returned to the shipyard after her six-day Navy acceptance trials in the Atlantic. A giant broom was affixed to her masthead to signify a clean sweep of the trials. Capt. W. M. Ryan, President of the Naval Board of Inspection and Survey, stated:

"The ship generally performed in an excellent manner. The cleanliness and upkeep were outstanding. The fine workmanship throughout the ship reflects great credit upon all hands concerned with its building. Like all new ships there are bugs which must be worked out, but we feel that there is nothing that cannot be overcome."

Chapter 13

The Evolution of Aircraft Carriers

USS Constellation (CVA-64), sister ship to Kitty Hawk, fires a Terrier missile. Basic design of the Kitty Hawks is of the Forrestal class. The angled area of flight deck is some 40 feet longer than the Forrestal's, and catapults and elevators installed have greater capacity.
Source: www.history.navy.mil

The plant for the nuclear-powered aircraft carrier was designed under supervision of V.Adm., Hyman Rickover.

Designated CVA(N)-65, *Enterprise* was commissioned on November 25, 1961, at Norfolk, with Capt. V. P. dePoix commanding. The world's first nuclear-powered carrier has a length of 1040 feet between perpendiculars and an extreme breadth of 252 on the flight deck. Each of the four deck edge elevators covers about 4000 square feet. *Enterprise* is the first carrier to have elevators for pilots in lieu of escalators. She displaces 85,350 tons.

Dominating feature of Enterprise's silhouette is her box-like island. Note the size of the sailors lining the deck as compared to the huge island. Source: instapinch.com

The communications equipment on the carrier is believed to be the largest assortment ever assembled on any ship. Besides more than 1800 telephones, there is

Chapter 13

The Evolution of Aircraft Carriers

the complexity of numerous radio circuits, teletypes, a pneumatic tube arrangement to carry messages from one station to another, and numerous announcing systems, several of which have speakers throughout the ship. She is the first ship of the U.S. Navy's Atlantic Fleet to have the Navy Tactical Data System installed.

USS America (CVA-66) Underway on 31 August 1965. Aircraft parked on her flight deck include nine A-4 "Skyhawk" attack planes, four F-4 "Phantom II" fighters and three A-5A "Vigilante" heavy attack planes. With angled decks, four powerful steam catapults, multiple starboard edge elevators, and tremendously powerful engines this new class of aircraft carriers were like independent fighting machines that could bring the full force of the United States and her Allies to bear around the world
Source: www.history.navy.mil

The *Enterprise* is equipped with four type C-13 steam driven catapults with an energy potential of 60,000,000 foot-pounds. With this power, an aircraft weighing 78,000 pounds can be accelerated to 160 mph from a standing start, in a distance of 250 feet. All of the aircraft aboard can be launched at the rate of one every 15 seconds while using all four cats.

The size of *Enterprise's* island structure was dictated by the size of the two radar screens that flank each of its four sides. This newly developed radar system is the most powerful to be installed on a floating platform, according to Capt. de Poix. Its far-reaching, three-dimensional capability is enhanced by its height above the water line. The silhouette is distinctive.

"Propulsion and control characteristics of the ship offer great tactical flexibility," said Capt. de Poix in mid-1962. **"There are four rudders, one almost directly astern of each propeller. This provides excellent**

Chapter 13

The Evolution of Aircraft Carriers

maneuverability at all speeds as well as tactical diameters in turns which compare with much smaller ships."

"Her ability to launch a strike on the enemy from one position, recover, and launch another 24 hours later from an unpredictable position more than 800 miles away from her previous strike position will constantly be a factor in causing the enemy to utilize protective forces that could be deployed elsewhere."

"If a show of force is required, *Enterprise* can be on distant station in a shorter period of time than any other ship in the Fleet."

Appendix I

Naval - Aviation Chronology in World War II

Appendix I

Naval - Aviation Chronology in World War II

World War II 1940-1945

Thirty years after the Navy had acquired its first airplane, and only 19 years after it had acquired its first aircraft carrier, naval aviation faced the supreme test of war. When it was called upon to carry the fight to the enemy, it not only carried out its tasks, but forged ahead to become the very backbone of fleet striking power.

If it had not already been shown in combat before the United States entered the war, all doubts as to the potency of naval air power were removed by the infamous, yet skillfully executed attack on Pearl Harbor, when Japanese carrier aircraft in one swift stroke devastated our principal naval base in the Pacific and eliminated a major portion of the Navy's heavy surface power. That our own forces had the kernel of a similar potential was demonstrated on a much smaller scale as carrier forces struck the first retaliatory blows.

The geographic position of the United States put it squarely between two wars that had little in common. Air operations on the Atlantic side, except for participation in three amphibious operations, were essentially a blockade and a campaign to protect ships delivering raw materials to our factories and war munitions and reinforcements to our Allies. In the Pacific, it was a matter of stopping an enemy advance which, in a few short months, had spread over all the western and parts of the south and central Pacific, and then carrying out the bitterly contested task of driving him homeward across the broad expanse of an island-dotted sea.

The country was hardly ready for either campaign. The Navy and Marine Corps air arms could muster only 7 large and 1 small aircraft carriers, 5 patrol wings and 2 Marine aircraft wings, 5,900 pilots and 21,678 enlisted men; 5,233 aircraft of all types including trainers, and a few advanced air bases. But aided by its distance from the enemy and fortunate in its industrial power, the United States built the ships, planes and equipment, and trained the land, sea and air forces that ultimately beat down the enemy, drove him from strategically located bases, cut off his raw materials, and placed the allied forces in position to launch final air and amphibious offensives against his homeland--offensives that were made unnecessary as the awesome destructive power of the atom was released upon Hiroshima and Nagasaki.

Appendix I

The Evolution of Aircraft Carriers

For the first time in history, naval engagements were fought entirely in the air without opposing surface forces sighting each other. New words and phrases entered the aviator's lexicon; words like air support, hunter-killer, JATO, CIC, CAP, bogie, scramble and splash. Radar lighted the night and gave new eyes to the fleet; advances in technology, particularly in electronics, improved the defense and added power to the offense. The scientist contributed directly to the war effort in both the development of specialized equipment and in the application of scientific principles to operational tactics. Logistics took on new importance. Refueling and replenishment at sea were developed to a high art and increased the mobility and staying power of fleet forces.

In the course of the war, Navy and Marine pilots destroyed over 15,000 enemy aircraft in the air and on the ground, sank 174 Japanese warships, including 13 submarines, totaling 746,000 tons, sank 447 Japanese merchant ships totaling 1,600,000 tons and, in the Atlantic, destroyed 63 German U-boats. (In combination with other agents, Navy and Marine air helped sink another 157,000 tons of war and 200,000 tons of merchant ships and another six Japanese and 20 German submarines.) It was a creditable record, but the Navy's air arm did not play an entirely independent role. It operated as it had developed, as an integral part of naval forces, contributing its full share to the power of the fleet and to the achievement of its mission in controlling the sea.

Many have said that World War II witnessed the full development of aviation, but generalities are often misleading. Many of the opinions expressed before the war on the effect of air power on naval operations were shown up as misconceived, if not false, theories. The bombing tests of the 1920's proved to some that navies were obsolete and that no ship could again operate within the range of land-based air, but carrier task force operations in the war gave a lie to such conclusions. Advocates of independent air power questioned both the possibility and the usefulness of close air support for troops, but such support was proved not only possible but indispensable. Those who questioned the importance of the airplane to navies were equally off the mark. The disappointment of naval officers who visualized fleet engagements in the tradition of Trafalgar and Jutland was no doubt as great as that of the air power theorists who had seen their predictions go awry. By test of war it had become exceedingly clear that neither an Army nor a Navy could either survive or achieve an objective in war without first achieving superiority in the air. It had also become clear that neither could exert as much force by itself as it could with the aid of air striking power. Aviation had indeed come of age.

Appendix I

The Evolution of Aircraft Carriers

1940

JANUARY

4--Project Baker was established in Patrol Wing 1 for the purpose of conducting experiments with blind landing equipment.

FEBRUARY

15--The Commander in Chief, U.S. Fleet, noting that reports on air operations in the European War stressed the need of reducing aircraft vulnerability, recommended that naval aircraft be equipped with leak-proof or self-sealing fuel tanks and with armor for pilots and observers. Although the Bureaus of Aeronautics and Ordnance had been investigating these forms of protection for two years, this formal statement of need gave added impetus and accelerated procurement and installation of both armor and self-sealing fuel tanks.

24--The Bureau of Aeronautics issued a contract for television equipment, including camera, transmitter, and receiver, that was capable of airborne operation. Such equipment promised to be useful both in transmitting instrument readings obtained from radio-controlled structural flight tests, and in providing target and guidance information necessary should radio-controlled aircraft be converted to offensive weapons.

27--Development of the "Flying Flapjack", a fighter aircraft with an almost circular wing, was initiated with notice of a contract award to Vought-Sikorsky Aircraft for the design of the V-173--a full-scale flying model (as distinguished from a military prototype). This design, based upon the research of a former NACA engineer, Charles H. Zimmerman, was attractive because it promised to combine a high speed of near 500 m.p.h. with a very low takeoff speed.

29--The Bureau of Aeronautics initiated action that led to a contract with Professor H. O. Croft, University of Iowa, to investigate the possibilities of a turbojet propulsion unit for aircraft.

MARCH

19--To assist in the identification of U.S. aircraft on the Neutrality Patrol, Fleet activities were authorized to apply additional National Star Insignia on the sides of the fuselage or hull of aircraft so employed.

Appendix I

The Evolution of Aircraft Carriers

22--Development of guided missiles was initiated at the Naval Aircraft Factory with the establishment of a project for adapting radio controls to a torpedo-carrying TG-2 airplane.

APRIL

23--Commander D. Royce was designated to represent the Navy on an Army Air Corps Evaluation Board for rotary-wing aircraft. This board was established incidental to legislation directing the War Department to undertake governmental development of rotary-wing aircraft.

25--*Wasp* was commissioned at Boston, Captain J. W. Reeves, Jr., commanding.

MAY

20--The Commanding Officer of the destroyer Noa (DD 343) reported on successful operations conducted off the Delaware Capes in which an XSOC-1, piloted by Lieutenant G. L. Heap, was hoisted over the side for takeoff and was recovered by the ship while underway. As an epilogue to preliminary operations conducted at anchor on 15 May, Lieutenant Heap made an emergency flight transferring a stricken seaman from the Noa in Harbor of Refuge, Del., to the Naval Hospital, Philadelphia.

27--The Secretary directed that six destroyers of the DD 445-Class be equipped with catapult, plane, and plane handling equipment. DD's 476-481, *Pringle, Stanly, Hutchins, Stevens, Halford*, and *Leutze*, were subsequently selected. Shortcomings in the plane hoisting gear led to removal of the aviation equipment from the first three ships prior to their joining the fleet in early 1943. In October 1943, after limited aircraft operations by the *Stevens* and the *Halford*, aviation equipment was ordered removed from them and plans for its installation on the Leutze were canceled.

JUNE

14--The Naval Expansion Act included authorization for an increase in aircraft carrier tonnage of 79,500 tons over the limits set 17 May 1938, and a revision of authorized aircraft strength to 4,500 useful airplanes.

15--Congress revised its previous action and set the aircraft ceiling at 10,000 useful airplanes, including 850 for the Naval Reserve, and not more than 48 useful airships.

Appendix I

The Evolution of Aircraft Carriers

25--The Aeronautical Engineering Duty Only (AEDO) designation was abolished and all men appointed to that special duty were designated for Engineering Duty Only (EDO).

25--The Chief of Naval Operations promulgated plans for an expanded flight training program calling for the assignment of 150 students per month beginning 1 July, and a regular increase to an entry rate of 300 per month within a year.

27--The President established a National Defense Research Committee to correlate and support scientific research on the mechanisms and devices of war. Among its members were officers of the War and Navy Departments appointed by the respective Secretaries. Although research on the problems of flight was specifically excluded from its functions, this organization made substantial contributions in various fields of importance to naval aviation, including airborne radar.

JULY

14--The initial meeting of what became the National Defense Research Committee's Division 14, or Radar Division, was attended by Alfred L. Loomis, Ralph Bowen, E. L. Bowles and Hugh H. Willis. In this and subsequent meetings with other scientists, this group defined its mission as "to obtain the most effective military application of microwaves in minimum time." In carrying out this mission, Division 14 developed airborne radar used in the Navy for aircraft interception, airborne early warning and other more specialized applications.

19--Authorization for a further expansion of the Navy provided an increase of 200,000 tons in the aircraft carrier limits set the previous month, and a new aircraft ceiling of 15,000 useful planes. The act also allowed further increases in aircraft strength on Presidential approval.

AUGUST

5--The Chief Of Naval Operations established general ground rules for exchange of scientific and technical information with a British mission, generally known as the Tizard Mission after its senior member Sir Henry Tizard. In general, free exchange of information was expected on matters concerning aviation, including the field later called radar. The degree of exchange actually achieved surpassed expectations so that the coming of the Tizard Mission served as a benchmark in the interchange of scientific and technical information regarding World War II weaponry.

Appendix I

The Evolution of Aircraft Carriers

12--The Bureau of Ordnance informally requested the National Defense Research Committee to sponsor development, on a priority basis, of proximity fuzes with particular emphasis on anti-aircraft use. Such fuzes had been under consideration for some time and the decision to undertake development followed receipt from the Tizard Mission of reports of British progress.

17--Section T (so called for its Chairman, Dr. Merle A. Tuve) of Division A, National Defense Research Committee, was established to examine the feasibility of various approaches to developing a proximity fuze. Eight days later, a contract was issued to the Department of Terrestrial Magnetism, Carnegie Institution of Washington, for the research that culminated in the radio VT fuze for anti-aircraft guns and both radio and photoelectric VT fuzes for bombs and rockets.

29--The exchange with the British Tizard Mission of scientific and technical information concerning radar began at a conference attended by Sir Henry Tizard, two of his associates, and representatives of the U.S. Army and Navy including Lieutenant J.A. Moreno of the Bureau of Aeronautics. The initial conference dealt primarily with the British techniques for detecting German bombers but touched upon means of identifying friendly aircraft. In follow-on meetings, British developments of shipboard and airborne radar were also discussed. A British disclosure growing out of this exchange of particular importance for airborne radar application was the cavity magnetron, a tube capable of generating high power radio waves of a few centimeters in length.

SEPTEMBER

2--In exchange for 50 four-stack destroyers, Great Britain, by formal agreement ceded to the United States for a period of 99 years, sites for naval and air bases in the Bahamas, Jamaica, St. Lucia, Trinidad, Antigua, and British Guiana, and extended similar rights freely and without consideration for bases in Bermuda and Newfoundland. Acquisition of these sites advanced our sea frontiers several hundred miles and provided bases from which naval ships and aircraft could cover strategically important sea approaches to our coast and to the Panama Canal.

OCTOBER

3--The Chief of Naval Operations requested the Naval Attache at London to obtain samples of a variety of British radio echo equipments (radar) including aircraft installations for interception (AI), surface vessel detection (ASV) and aircraft identification (IFF).

Appendix I

The Evolution of Aircraft Carriers

5--The Secretary of the Navy placed all divisions and aviation squadrons of the Organized Reserve on short notice for call to active duty and granted authority to call Fleet Reservists as necessary. On the 24th the Bureau of Navigation announced plans for mobilizing the aviation squadrons, which called for one third to be ordered to active duty by 7 November and all by 1 January 1941.

9--The Secretary approved a recommendation by the General Board, that 24 of the authorized submarines be equipped to carry aviation gasoline for delivery to seaplanes on the water. This was in addition to Nautilus (SS 168) which had demonstrated her ability to refuel patrol planes and had conducted a successful test dive to 300 feet with aviation gasoline aboard; and to Narwhal (SC 1) and Argonaut (SF 7) which were being altered to carry 19,000 gallons of aviation gasoline each.

11--The Technical Aide to the Secretary of the Navy, Rear Admiral Harold G. Bowen, proposed a program for development of radio ranging equipment (radar) which formed the basis for the Navy's pre-war development program. In addition to identification equipment and ship-based radar, this program included an airborne radar for surface search.

23--Within the Atlantic Squadron, an administrative command was set up for carrier aviation entitled, "Aircraft, Atlantic Squadron."

24--An administrative command for patrol aviation in the Atlantic Squadron was set up under the title, "Patrol Wings, Atlantic Squadron."

28--The Chief of Naval Operations reported that aircraft with some form of armor and fuel protection were just beginning to go into service use, and that within a year all fleet aircraft, except those assigned Patrol Wing 2, would have such protection.

NOVEMBER

1--A reorganization of the Fleet changed the administrative organization of aviation by dividing the forces between two oceans which was the beginning of the independent development of forces according to strategic requirements. In the Atlantic, aviation was transferred from Scouting Force to Patrol Force, which was formed in place of the Atlantic Squadron as a fleet command parallel to Scouting Force, and set up under Commander Aircraft Patrol Force and Commander Patrol Wings Patrol Force. In the Pacific, Patrol Wings remained attached to Scouting Force under the combined command Commander Patrol Wings U.S. Fleet and Commander Aircraft Scouting Force.

Appendix I

The Evolution of Aircraft Carriers

11--The first general meeting of the Radiation Laboratory was held at the Massachusetts Institute of Technology. The Radiation Laboratory, as principal scientific and developmental agency of Division 14 of NDRC, was to become instrumental in many aspects of airborne radar development.

15--The seaplane tender Curtiss first of two ships of her class, was commissioned at Philadelphia, Commander S. P. Ginder commanding.

15--Naval air operations began from Bermuda. First to operate were the planes of Patrol Squadron 54 based on George E. Badger.

16--The Bureau of Aeronautics established a catapult procurement program for Essex class carriers. One flight deck catapult and one athwartships hangar deck catapult were to be installed on each of 11 carriers.

18--The Chief of Naval Operations authorized use of the abbreviation, "RADAR", in unclassified correspondence and conversation and directed that the phrase, Radio Detection and Ranging Equipment, be used in lieu of terms such as Radio Ranging Equipment, Radio Detection Equipment, Radio Echo Equipment, or Pulse Radio Equipment.

DECEMBER

30--The Bureau of Aeronautics directed that fleet aircraft be painted in non-specular colors. Ship-based aircraft were to be light gray all over; patrol planes were to be light gray except for surfaces seen from above which were to be blue gray.

1941

FEBRUARY

1--The Atlantic and Pacific Fleets were established, completing the division begun the previous November and changing the titles of aviation commands in the Atlantic Fleet to "Aircraft, Atlantic Fleet" and "Patrol Wings, Atlantic Fleet." No change was made in the Pacific Fleet aviation organization at this time.

10--As an initial step in training patrol plane pilots to make blind landings, using radio instrument landing equipment which was being procured for all patrol aircraft and their bases, a one-month course of instruction began under Project Baker. This was attended by one pilot from each of 13 squadrons; by one radioman from each of five patrol wings; and by two radiomen from each of five Naval Air Stations.

26--An extensive modification of aircraft markings added National Star Insignia to both sides of the fuselage or hull and eliminated those on the upper right and lower left wings; discontinued the use of colored tail markings, fuselage bands and cowl markings; made removal of vertical red, white and blue rudder stripes mandatory; and changed the color of all markings, except the National Insignia, to those of least contrast to the background.

MARCH

1--Support Force, Atlantic Fleet, was established for operations on the convoy routes across the North Atlantic. Its component patrol squadrons were placed under a Patrol Wing established at the same time.

11-The President was empowered by an Act of Congress to provide goods and services to those nations whose defense he deemed vital to the defense of the United States, thus initiating a Lend-Lease program under which large quantities of the munitions and implements of war were delivered to our allies. The *Archer* (BAVG 1) transferred on 17 November 1941, as the first of 38 escort carriers transferred to the United Kingdom during the war.

17--The Chief of the Bureau of Aeronautics approved a proposal for establishing a special NACA committee to promptly review the status of jet propulsion and recommend plans for its application to flight and assisted takeoff.

28--The Commanding Officer of *Yorktown* after five months operational experience with the CXAM radar, reported that aircraft had been tracked at a distance of 100 miles and recommended that friendly aircraft be equipped with electronic identification devices and carriers be equipped with separate and complete facilities for tracking and plotting all radar targets.

APRIL

19--Development of a Glomb (Glider Bomb) of a guided missile was initiated at the Naval Aircraft Factory. The Glomb was a glider designed to be towed long distances by a powered aircraft, released in the vicinity of the target, and guided by radio control in its attack. It was equipped with a television camera to transmit a view of the target to the control plane.

20--The first successful test of electronic components of a radio-proximity fuze was made at a farm in Vienna, Va., as a radio oscillator, or sonde, which had

Appendix I

The Evolution of Aircraft Carriers

been fired from a 37-mm pack howitzer, made radio transmissions during its flight. The demonstration, that radio tubes and batteries could be constructed sufficiently rugged to withstand firing from a gun, led Section T of the National Defense Research Committee to concentrate upon the radio-proximity fuze for anti-aircraft guns.

26--The Naval Aircraft Factory project officer reported that an unmanned O3U-6 airplane under radio control had been successfully flight-tested beyond the safe bounds of piloted flight and that the information thus obtained had been of great value in overcoming flutter encountered at various speeds and accelerations.

28--*Pocomoke*, first of two seaplane tenders of her class, was commissioned at Portsmouth, Va., Commander J.D. Price commanding.

30--In the initial step towards establishing a glider development program, the Naval Aircraft Factory was requested to undertake preliminary design of a personnel and equipment transport glider. As work progressed and requirements were further clarified, development was initiated for 12- and 24-place amphibian gliders constructed of wood or plastic by firms not already engaged in building military aircraft.

30--The Commanding Officer NAS Lakehurst directed that the metal-clad airship, ZMC-2, be salvaged and the car complete with engines, instruments and appurtenances be assigned to the Lighter-Than-Air Ground School at Lakehurst. The ZMC-2, completed in August 1929, had been flown over 2,250 hours.

MAY

2--Fleet Air Photographic Unit, Pacific, was established under Commander Aircraft Battle Force, preceding by one day the establishment of a similar unit in the Atlantic Fleet under Commander Patrol Wings, Atlantic.

3--Project Roger was established at the Naval Aircraft Factory to install and test airborne radar equipment. Its principal assignment involved support of the Radiation Laboratory at the Massachusetts Institute of Technology and the Naval Research Laboratory in various radar applications including search and blind bombing and in radio control of aircraft.

8--The establishment of Aviation Repair Units 1 and 2 was directed to provide a nucleus of aircraft repair and maintenance personnel ready for overseas deployment as advanced bases were established.

Appendix I

The Evolution of Aircraft Carriers

10--The Naval Aircraft Factory reported that it was negotiating with the Radio Corporation of America for the development of a radio altimeter suitable for use in radio-controlled assault drones.

15--The seaplane tender *Albemarle* arrived at Argentia, Newfoundland, to establish a base for Patrol Wing, Support Force operations and to prepare for the imminent arrival of VP-52, the first squadron to fly patrols over the North Atlantic convoy routes.

21--The Bureau of Aeronautics requested the Engineering Experiment Station, Annapolis, Md., to undertake development of a liquid-fueled assisted takeoff unit for use on patrol planes. This marked the Navy's entry into the field that later came to be called JATO, and was the Navy's first development program, other than jet exhaust from reciprocating engines, directed towards utilizing jet reaction for aircraft propulsion.

27--The President proclaimed that an unlimited national emergency confronted the country, requiring that its military, naval, air, and civilian defenses be put on the basis of readiness to repel any and all acts or threats of aggression directed toward any part of the Western Hemisphere.

JUNE

2--*Long Island*, first escort carrier of the U.S. Navy, was commissioned at Newport News, VA., Commander D.B. Duncan commanding. Originally designated AVG 1, the *Long Island* was a flush-deck carrier converted in 67 working days from the cargo ship Mormacmail.

4--The Naval Aircraft Factory reported that development of airborne television had progressed to the point that signals transmitted by this means could be used to alter the course of the transmitting plane.

11--An Aircraft Armament Unit was formed at NAS Norfolk with Lieutenant Commander W.V. Davis as Officer in Charge, to test and evaluate armament installations of increasing complexity.

28--To strengthen the provisions for utilizing science in war, the President created the Office of Scientific Research and Development and included in its organization the National Defense Research Committee and a newly established Committee on Medical Research.

Appendix I

The Evolution of Aircraft Carriers

30--Turboprop engine development was initiated as a joint Army-Navy project, with a Navy contract to Northrop Aircraft for the design of an aircraft gas turbine developing 2,500 hp. at a weight of less than 3,215 pounds.

JULY

1--The first landing, takeoff, and catapult launching from an escort carrier were made aboard Long Island, by Lieutenant Commander W.D. Anderson, Commanding Officer of VS-201.

1--The Test, Acceptance and Indoctrination Units that had been established at San Diego and Norfolk in May to fit out new patrol aircraft and to indoctrinate new crews in their use, were expanded and set up as separate commands. The San Diego Unit, which retained its original name, was placed under Commander Aircraft Scouting Force, and the Norfolk unit became Operational Training Squadron under Commander Patrol Wings, Atlantic.

1--Patrol Wing, Support Force, was redesignated and established as Patrol Wing 7, Captain H. M. Mullinnix commanding.

3--The Seaplane tender *Barnegat*, first of 26 ships of her class was commissioned at Bremerton, Commander F. L. Baker commanding.

4--Planes of Patrol Squadron 72, based on *Goldsborough*, flew protective patrols from Reykjavik, Iceland, until the 17th, to cover the arrival of Marine Corps garrison units from the United States.

7--The First Marine Aircraft Wing, composed of a Headquarters Squadron and Marine Air Group 1, was organized at Quantico under command of Lieutenant Colonel Louis E. Woods. It was first of its type in the Marine Corps and the first of five wings organized during the war period.

8--Patrol Wing 8 was established at Norfolk, Commander J.D. Price commanding.

12--The Naval Research Laboratory was transferred from the Office of the Secretary to the cognizance of the Bureau of Ships, and a Naval Research and Development Board was established in the Office of the Secretary composed of representatives of the Chief of Naval Operations and the Bureaus of Aeronautics, Ordnance, Ships, and Yards and Docks, and led by a civilian scientist with the title Coordinator of Research and Development. Dr. J.C. Hunsaker served as coordinator until December when he was relieved by Rear Admiral J.A. Furer.

Appendix I

The Evolution of Aircraft Carriers

17--The organization for development of proximity fuzes was realigned so that Section T could devote its entire effort to radio-proximity fuzes for anti-aircraft projectiles. Responsibility for photoelectric and radio fuzes for bombs and rockets was transferred to Section E of the National Defense Research Committee at the National Bureau of Standards.

18--Commander J. V. Carney, Senior Support Force Staff Officer, reported that British type ASV radar has been installed in one PBY-5 each of VP-71, VP-72, and VP-73 and two PBM-1's of VP-74. Initial installation of identification equipment (IFF) was made about the same time. In mid-September radar was issued for five additional PBM-1's of VP-74 and one PBY-5 of VP-71, and shortly thereafter for other aircraft in Patrol Wing 7 squadrons. Thereby the Wing became the first operational unit of the U.S. Navy to be supplied with radar-equipped aircraft. Its squadrons operated from Norfolk, Quonset Point and advanced bases on Greenland, Newfoundland and Iceland during the last months of the neutrality patrol.

18--Aviation was given representation on the highest of the Army and Navy boards as membership of the Joint Board was revised to include the Deputy Chief of Staff for Air and the Chief of the Bureau of Aeronautics.

21--The requirement that all students assigned to the carrier-plane phase of flight training be given time in each of the three basic aircraft types was abolished, and the practice of assigning students to specialized training in either fighters, scout bombers or torpedo planes began.

25--Thirty P-40's and three primary training planes of the 33rd Pursuit Squadron, Army Air Forces, were loaded aboard *Wasp* at Norfolk for transport to Reykjavik, Iceland.

28--To establish a continuing organization for training flight crews, the Chief of Naval Operations directed that action be taken as expeditiously as practicable to provide additional gunnery and tactical training in the pilot training program; to establish within the Atlantic and Pacific Fleets at Norfolk and San Diego, Advanced Carrier Training groups to indoctrinate newly designated Naval Aviators in the operation of current model carrier aircraft; and to assign a number of patrol squadrons in each fleet the primary task of providing familiarization, indoctrination, advanced gunnery and tactical training for new flight crews.

28--The Operational Training Squadron of the Atlantic Fleet, and the Test, Acceptance and Indoctrination Unit of the Pacific Fleet were redesignated Transition Training Squadron, Atlantic and Pacific respectively.

Appendix I

The Evolution of Aircraft Carriers

29--The Secretary of the Navy approved the installation of a Radar Plot aboard carriers as "the brain of the organization" protecting the fleet from air attack. The first installation was planned for the island structure of the *Hornet*.

AUGUST

1--A Microwave (AI-10) radar developed by the Radiation Laboratory and featuring a Plan Position Indicator (or PPI) was given its initial airborne test in the XJO-3 at Boston Airport. During the test flights, which continued through 16 October, Radiation Laboratory scientists operated the radar and devised modifications while naval personnel from Project Roger (usually Chief Aviation Pilot C. L. Kullberg) piloted the aircraft. During the tests, surface vessels were detected at ranges up to 40 miles; radar-guided approaches against simulated enemy aircraft were achieved at ranges up to 3.5 miles. Operational radars which were developed from this equipment were capable of searching a circular area and included the ASG for K-type airships and the AN/APS-2 for patrol planes.

1--The Bureau of Aeronautics requested the Naval Research Laboratory to develop radar guidance equipment for assault drones, both to relay target information to a control operator and to serve as automatic homing equipment. This marked the initiation of radar applications to guided missiles.

6--Patrol Squadrons 73 and 74 initiated routine air patrols from Reykjavik, Iceland, over North Atlantic convoy routes.

6--In recognition of the radical change which radar was causing in the method of using fighters to protect the fleet, the Chief of Naval Operations issued a "Tentative Doctrine for Fighter Direction from Aircraft Carriers" and directed that carriers and other ships equipped with radar immediately organize fighter direction centers.

7--The Chief, Bureau of Aeronautics issued a preliminary plan for installing radar in naval aircraft. Long range search radar (British ASV or American ASA) was to be installed in patrol planes. Short range search radar (British Mk II ASV modified for Fleet Air Arm or American ASB) was to be installed in one torpedo plane in each section commencing with the TBF while space needed for search radar was to be reserved in new scout-dive-bombers and scout-observation planes. Interception equipment, when available, would be installed in some F4U's and a British AI Mk IV radar was being installed in an SBD with a view to its use as an interim interceptor. The plan also included installation of appropriate radio altimeters in patrol and torpedo planes, and recognition equipment in all service airplanes.

Appendix I

The Evolution of Aircraft Carriers

SEPTEMBER

5--Artemus L. Gates, Naval Aviator No. 65 and member of the First Yale Unit of World War I, took the oath of office as Assistant Secretary of the Navy for Aeronautics; the first to hold the office since the resignation of David S. Ingalls in 1932.

9--The Bureau of Aeronautics requested the National Defense Research Committee and the Naval Research Laboratory to develop an interceptor radar suitable for installation in single engine, single seat fighters such as the F4U.

OCTOBER

1--The Aviation Supply Office was established at Philadelphia, under the joint cognizance of the Bureau of Aeronautics and the Bureau of Supplies and Accounts, to provide centralized control over the procurement and distribution of all aeronautical materials regularly maintained in the general stock.

8--Organizational provision for guided missiles was made in the fleet by the establishment of "Special Project Dog" in Utility Squadron 5, to test and operate radio-controlled offensive weapons and to train personnel in their use. VJ-5 was also directed to develop a radio-controlled fighter plane--"aerial ram" or "aerial torpedo"--to be flown into enemy bomber formations and exploded.

13--The Bureau of Aeronautics directed that all fleet aircraft be painted non-specular light gray except for surfaces seen above which were to be blue-gray. In late December, this color scheme was extended to shore based airplanes except trainers.

20--Hornet was commissioned at Norfolk, Captain Marc A. Mitscher commanding.

21--In tests with MAD gear (Magnetic Airborne Detector), a PBY from NAS Quonset Point, located the submarine S-48. The tests were carried out in cooperation with the National Defense Research Committee.

29--Patrol Squadron 82 received the first of a planned full complement of PBO-1's at NAS Norfolk. Assignment of these aircraft, actually destined for the British and painted with British markings, was the beginning of what became an extensive use of land planes by patrol squadrons during the war and, although it was not yet apparent, was the first move toward the eventual elimination of the flying boat from patrol aviation.

Appendix I

The Evolution of Aircraft Carriers

NOVEMBER

1--By Executive Order the President directed that, until further orders, the Coast Guard operate as a part of the Navy subject to the orders of the Secretary of the Navy.

18--Doctor L. A. DuBridge of the Radiation Laboratory reported that the initial design of a 3-cm aircraft intercept radar was completed.

26--Kitty Hawk, first of two aircraft ferries, was commissioned, Commander C.E. Rogers commanding.

DECEMBER

1--Patrol Wing 9 began forming at Quonset Point with Lieutenant Commander T. U. Sisson as prospective Commanding Officer.

7--Japanese carrier aircraft launched a devastating attack on ships at Pearl Harbor and on the military and air installations in the area. The three aircraft carriers of the Pacific Fleet were not present. *Saratoga*, just out of overhaul, was moored at San Diego. *Lexington* was at sea about 425 miles southeast of Midway toward which she was headed to deliver a Marine Scout Bombing Squadron. *Enterprise* was also at sea about 200 miles west of Pearl Harbor, returning from Wake Island after delivering a Marine Fighter Squadron there. Her Scouting Squadron 6, launched early in the morning to land at Ewa Airfield, arrived during the attack and engaged enemy aircraft.

9--The Secretary of the Navy authorized the Bureau of Ships to contract with the RCA Manufacturing Company for a service test quantity of 25 sets of ASB airborne search radar. This radar had been developed by the Naval Research Laboratory (under the designation XAT) for installation in dive bombers and torpedo planes.

10--Aircraft from Enterprise attacked and sank the Japanese submarine *I-70* in waters north of the Hawaiian Islands. This was one of the submarines used to scout the Hawaiian area in connection with the Pearl Harbor attack and the first Japanese combatant ship sunk by United States aircraft during World War II.

10--Antisubmarine patrols over the South Atlantic were initiated by Patrol Squadron 52, equipped with Catalinas operating from Natal, Brazil.

12--The Naval Air Transport Service (NATS) was established under the Chief of Naval Operations to provide rapid air delivery of critical equipment, spare

Appendix I

The Evolution of Aircraft Carriers

parts, and specialist personnel to naval activities and fleet forces all over the world.

14--Patrol Wing 10 departed Cavite and, with its two patrol squadrons and four seaplane tenders, began withdrawal from the Philippines. Before reaching Australia it operated from various bases along the way, including Balikpapan, Soerabaja, and Ambon in the Netherlands East Indies.

15--Patrol Wing 8 transferred from Norfolk to Alameda for duty on the west coast.

16--The Secretary of the Navy approved an expansion of the pilot training program from the existing schedule of assigning 800 students per month to one calling for 2,500 per month thereby leading to a production of 20,000 pilots annually by mid-1943.

17--The Naval Research Laboratory reported that flight tests in a PBY of radar utilizing a duplexing antenna switch had been conducted with satisfactory results. The duplexing switch made it possible to use a single antenna for both transmission of the radar pulse and reception of its echo; thereby, the necessity for cumbersome "yagi" antenna no longer existed, a factor which contributed substantially to the reliability, and hence the effectiveness, of World War II airborne radar.

17--Seventeen SB2U-3 Vindicators of VMSB-231, led by a PBY of Patrol Wing 1, arrived at Midway Island from Oahu, completing the longest mass flight by single-engine aircraft then on record in 9 hours, 45 minutes. It was the same squadron that was en route to Midway on 7 December aboard *Lexington* when reports of the attack on Pearl Harbor forced the carrier to turn back short of her goal.

18--Two-plane detachments from Patrol Wings 1 and 2, based in Hawaii, began scouting patrols from Johnston Island.

25--Two-plane detachments from squadrons at Pearl and Kaneohe began patrols from Palmyra Island, a principal staging base to the South Pacific.

1942

JANUARY

2--The first organized lighter-than-air units of World War II, Airship Patrol Group 1, Commander George H. Mills commanding, and Airship Squadron 12,

Appendix I

The Evolution of Aircraft Carriers

Lieutenant Commander Raymond F. Tyler commanding, were established at NAS Lakehurst.

5--A change in regulations, covering display of National Insignia on aircraft, returned the star to the upper right and lower left wing surfaces and revised rudder striping to 13 red and white horizontal stripes.

7--Expansion of naval aviation to 27,500 useful planes was approved by the President.

11--*Saratoga* while operating at sea 500 miles southwest of Oahu, was hit by a submarine torpedo and forced to retire for repairs.

11--Patrol Squadron 22, with PBY-5 Catalinas, joined Patrol Wing 10 at Ambon, the first aviation reinforcements from the Central Pacific to reach southwest Pacific Forces opposing the Japanese advance through the Netherlands East Indies.

14--The formation of four Carrier Aircraft Service Units (CASU) from four small Service Units, previously established in the Hawaiian area, was approved.

16--To protect the advance of Task Force 8 for its strike against the Marshall and Gilbert Islands, planes of Patrol Squadron 23 began daily search of the waters between their temporary base at Canton Island and Suva in the Fijis. These were the first combat patrols by aircraft in the South Pacific.

23--The first naval aircraft to operate in the Samoan Islands, OS2Us of VS-1-D14, arrived with Marine Corps reinforcements from San Diego.

29--Five-inch projectiles containing radio-proximity fuzes were test fired at the Naval Proving Ground, Dahlgren, and 52 percent of the fuzes functioned satisfactorily by proximity to water at the end of a 5-mile trajectory. This performance, obtained with samples selected to simulate a production lot, confirmed that the radioproximity fuze would greatly increase the effectiveness of anti-aircraft batteries and led to immediate small scale production of the fuze.

30--The Secretary authorized a glider program for the Marine Corps consisting of small and large types in sufficient numbers for the training and transportation of two battalions of 900 men each.

Appendix I

The Evolution of Aircraft Carriers

FEBRUARY

1--The Secretary of the Navy announced that all prospective naval aviators would begin their training with a three months' course emphasizing physical conditioning and conducted by Pre-Flight Schools to be established at universities in different parts of the country. The training began at the Universities of North Carolina and Iowa in May, the University of Georgia and St. Mary's College, Calif., in June, and at Del Monte, Calif., in January 1943.

1--First U.S. Carrier Offensive--Task Forces 8 (Vice Admiral W.F. Halsey) and 17 (Rear Admiral F.J. Fletcher), built around the carriers Enterprise and Yorktown, bombed and bombarded enemy installations on the islands of Wotje, Kwajalein, Jaluit, Makin, and Mili in the Marshall and Gilbert Islands.

12--The Chief of Naval Operations promulgated an advanced base program using the code names "Lion" and "Cub" to designate major and minor bases, and in July added "Oaks" and "Acorns" for aviation bases. This was the beginning of a concept of functional components which developed as the war progressed and which provided planners and commanders with a means of ordering standardized units of personnel, equipment, and material to meet any special need in any area, in much the same manner as ordering from a mail-order catalogue.

16--A Navy developed Air-Track blind landing system was in daily use in Iceland for landing flying boats. Other blind-landing systems were in various phases of development, and work on the Ground Controlled Approach system had progressed to the point that Navy personnel had made talk-down landings at the East Boston (Commonwealth) Airport.

17--The Commander in Chief, U.S. Fleet authorized removal of athwartships hangar deck catapults from *Wasp, Yorktown, Enterprise*, and *Hornet*.

21--The seaplane tender *Curtiss* and Patrol Squadron 14 arrived at Noumea, New Caledonia, to begin operations from what became a principal Navy base in the South Pacific during the first year of the war.

23--The Bureau of Aeronautics outlined a comprehensive program which became the basis for the wartime expansion of pilot training. In place of the existing 7-months course, the new program required 11 months for pilots of single or twin-engine aircraft and 12 months for four-engine pilots; and was divided into 3 months at Induction Centers, 3 months in Primary, 3 months in Intermediate and 2 or 3 months in Operational Training, depending on type aircraft used.

Appendix I

The Evolution of Aircraft Carriers

24--First Wake Island Raid--A striking force, (Vice Admiral W.F. Halsey) composed of the carrier *Enterprise* with cruiser and destroyer screen, attacked Wake Island.

26--The Navy's Coordinator of Research and Development requested the National Defense Research Committee to develop an expendable radio sonobuoy for use by lighter-than-air craft in antisubmarine warfare.

27--The seaplane tender Langley, formerly first carrier of the U.S. Navy, was sunk by enemy air attack 74 miles from her destination while ferrying 32 AAF P-40's to Tjilatjap, Java.

MARCH

1--Carrier Replacement Air Group 9 was established at NAS Norfolk under command of Commander William D. Anderson. It was the first numbered Air Group in the Navy and marked the end of the practice of naming air groups for the carriers to which they were assigned.

1--Ensign William Tepuni, USNR, piloting a Lockheed Hudson, PBO, of VP-82 based at Argentia, attacked and sank the *U-656* southwest of Newfoundland; the first German submarine sunk by U.S. forces in World War II.

2--Regularly scheduled operations by the Naval Air Transport Service were inaugurated with an R4D flight from Norfolk to Squantum.

4--First Raid on Marcus--*Enterprise*, as part of Task Force 16 (Vice Admiral W.F. Halsey), moved to within 1,000 miles of Japan to launch air attacks on Marcus Island.

7--Patrol Wing 10 completed withdrawal from the Philippines and the Netherlands East Indies, and established headquarters in Perth, for patrol operations along the west coast of Australia.

7--The practicability of using a radio sonobuoy in aerial anti-submarine warfare was demonstrated in an exercise conducted off New London by the K-5 blimp and the S-20 submarine. The buoy could detect the sound of the submerged submarine's propellers at distances up to three miles, and radio reception aboard the blimp was satisfactory up to five miles.

8--Inshore Patrol Squadron VS-2-D14, which had arrived at Bora Bora on 17 February, inaugurated air operations from the Society Islands.

Appendix I

The Evolution of Aircraft Carriers

9--VR-1, the first of 13 VR squadrons established under the Naval Air Transport Service during World War II, was established at Norfolk, Commander C.K. Wildman commanding.

10--A carrier air strike, launched from the Lexington and Yorktown in the Gulf of Papua, flew over the 15,000-foot Owen Stanley Mountains on the tip of New Guinea to hit Japanese shipping engaged in landing troops and supplies at Lae and Salamaua. One converted light cruiser, a large minesweeper, and a cargo ship were sunk and other ships damaged.

10--A contract with the Office of Scientific Research and Development became effective whereby the Johns Hopkins University agreed to operate a laboratory which became known as the Applied Physics Laboratory. This was one of several important steps in the transition of the radio-proximity fuze from development to large scale production. Other steps taken within the next 6 weeks included the organizational transfer of Section T from the National Defense Research Committee directly to the Office of Scientific Research and Development and the relocation of most of the Section T staff from the Carnegie Institution of Washington to the Applied Physics Laboratory at Silver Spring, Md.

26--Unity of command over Navy and Army air units, operating over the sea to protect shipping and conduct antisubmarine warfare, was vested in the Navy.

29--The forward echelon of Marine Fighter Squadron 212 arrived at Efate to construct an air strip from which the squadron initiated operations in the New Hebrides on 27 May.

APRIL

6--The administrative command Aircraft, Atlantic Fleet, was redesignated Carriers, Atlantic Fleet.

7--To provide aviation maintenance men with special training required to support air operations at advanced bases, Aircraft Repair Units 1 and 2 were merged to form the Advanced Base Aviation Training Unit (ABATU) at Norfolk.

9--A radio controlled TG-2 drone, directed by control pilot Lieutenant M. B. Taylor of Project Fox, made a torpedo attack on the destroyer Aaron Ward steaming at 15 knots in Narragansett Bay. Taylor utilized a view of the target obtained by a television camera mounted in the drone, and directed the attack so

Appendix I

The Evolution of Aircraft Carriers

that the torpedo was released about 300 feet directly astern of the target and passed under it.

10--A reorganization of the Pacific Fleet abolished the Battle and Scouting Forces and set up new type commands for ships and aviation. With the change, titles of the aviation type commands became Carriers, Pacific, and Patrol Wings, Pacific.

18--Raid on Tokyo--From a position at sea 668 miles from Tokyo, the carrier *Hornet* launched 16 B-25's of the 17th AAF Air Group led by Lieutenant Colonel J. H. Doolittle, USA, for the first attack on the Japanese homeland. The Hornet sortied from Alameda 2 April, made rendezvous with Enterprise and other ships of Task Force 16 (Vice Admiral W. F. Halsey) north of the Hawaiian Islands, and proceeded across the Pacific to the launching point without making port.

18--A night Fighter Development Unit was established to be located at NAS Quonset Point. This unit, originally named Project Argus was renamed Project Affirm to avoid confusion with the electronic element (Argus Unit) of an advanced base. Project Affirm's official purpose was to development and test night fighter equipment for Navy and Marine Corps aircraft; in addition, it developed tactics and trained officers and men for early night fighter squadrons and as night fighter directors.

19--Two tests of the feasibility of utilizing drone aircraft as guided missiles were conducted in Chesapeake Bay. In one, Utility Squadron VJ-5, utilizing visual direction, crash-dived a BG-1 drone into the water beyond its target, the wreck of San Marcos (LSD 25) and a live bomb exploder in the drone failed to detonate. The second and more successful test was conducted by Project Fox from CAA intermediate field, Lively, Va., using a BG-2 drone equipped with a television camera to provide a view of the target. Flying in a control plane 11 miles distant, Lieutenant M. B. Taylor directed the drones crash-dive into a raft being towed at a speed of 8 knots.

20--Wasp on special ferry duty out of Glasgow, Scotland, entered the Mediterranean and launched 47 Spitfires of the RAF to Malta. When the operation was duplicated on 9 May, it was the occasion for Winston Churchill's message, "Who says a Wasp cannot sting twice?"

24--A new specification for color of naval aircraft went into effect. The color of service aircraft remained non-specular light gray with non-specular blue-gray on surfaces visible from above. Advanced trainers were to be finished in glossy aircraft gray with glossy orange yellow on wing and aileron surfaces visible from above while primary trainers were to be finished glossy orange-yellow with gray landing gear.

30--The Air Operational Training Command was established with headquarters at Jacksonville, Fla. Four days later the Naval Air Stations at Jacksonville, Miami, Key West, and Banana River and their satellite fields were assigned to the new command.

MAY

4-8--Battle of the Coral Sea--In the first naval engagement of history fought without opposing ships making contact, United States carrier forces stopped a Japanese attempt to land at Port Moresby by turning back the covering carrier force. Task Force 17 (Rear Admiral F.J. Fletcher) with the carrier Yorktown, bombed Japanese transports engaged in landing troops in Tulagi Harbor, damaging several and sinking one destroyer (4 May); joined other Allied naval units including Task Force 11 (Rear Admiral A.W. Fitch) with the carrier Lexington south of the Louisiades (5 May); and after stationing an attack group in the probable track of the enemy transports, moved northward in search of the enemy covering force. Carrier aircraft located and sank the light carrier Shoho covering a convoy (7 May), while Japanese aircraft hit the separately operating attack group and sank one destroyer and one fleet tanker. The next day the Japanese covering force was located and taken under air attack, which damaged the carrier Shokaku. Almost simultaneously enemy carrier aircraft attacked Task Force 17, scoring hits which damaged *Yorktown* and set off uncontrollable fires on *Lexington*, as a result of which she was abandoned and was sunk (8 May). Although the score favored the Japanese, they retired from action and their occupation of Port Moresby by sea was deferred and finally abandoned.

10--The possibility of increasing the range of small aircraft, by operating them as towed gliders, was demonstrated at the Naval Aircraft Factory when Lieutenant Commanders W. H. McClure and R. W. Denbo hooked their F4F's to tow lines streamed behind a twin-engined BD (Army A-20), cut their engines and were towed for an hour at 180 knots at 7,000 feet.

10--*Ranger* on a transatlantic ferry trip reached a position off the African Gold Coast and launched 60 P-40 Warhawks of the Army Air Force to Accra, from which point they were flown in a series of hops to Karachi, India, for operations with the 10th AAF. This was the first of four ferry trips made by the *Ranger* to deliver AAF fighters across the Atlantic, the subsequent launches being accomplished on 19 July 1942, 19 January 1943, and 24 February 1943.

Appendix I

The Evolution of Aircraft Carriers

10--Inshore Patrol Squadron VS-4-D14 arrived in the Tonga Islands with the base construction and garrison convoy and set up facilities to conduct antisubmarine patrols from Nukualofa Harbor on Tongatabu.

11--The President ordered that an Air Medal be established for award to any person who, while serving in any capacity in or with the Army, Navy, Marine Corps, or Coast Guard after 8 September 1939, distinguishes or has distinguished himself by meritorious achievement while participating in aerial flight.

15--The design of the National Star Insignia was revised by eliminating the red disc in the center of the star, and use of horizontal red and white rudder striping was discontinued.

15--The Chief of Naval Operations ordered that an Assistant Chief of Naval Operations (Air) be established to deal with aviation matters directly under the Vice Chief of Naval Operations and that the Chief of the Bureau of Aeronautics fill the new office as additional duty. In complying with a further provision of the order that such readjustment of functions be made as would serve the interest of the order, the Vice Chief of Naval Operations subsequently concentrated the aviation functions already being performed in his office into a new Division of Aviation. The office was abolished in mid-June 1942.

15--A VR-2 flight from Alameda to Honolulu, the first transoceanic flight by NATS aircraft, initiated air transport service in the Pacific.

20--Rear Admiral J. S. McCain reported for duty as Commander Aircraft, South Pacific, a new command established to direct the operations of tender and shore-based aviation in the South Pacific area.

26--The feasibility of jet-assisted takeoff was demonstrated in a successful flight test of a Brewster F2A-3, piloted by Lieutenant (jg) C. Fink Fischer, at NAS Anacostia, using five British antiaircraft solid propellant rocket motors. The reduction in takeoff distance was 49 percent.

27--The transfer of Patrol Wing 4 from Seattle to the North Pacific began with the arrival of the Commander at Kodiak, Alaska.

JUNE

3-4--In an attempt to divert forces from the Midway area, a Japanese carrier force launched small raids on Dutch Harbor, hitting twice on the third and once on the fourth and doing considerable damage to installations ashore. PBY's

Appendix I

The Evolution of Aircraft Carriers

located the carriers on the fourth but attacks by 11th AAF bombers were unsuccessful.

3-6--The Battle of Midway--A strong Japanese thrust in the Central Pacific to occupy Midway Island, was led by a four- carrier Mobile Force, supported by heavy units of the Main Body (First Fleet) and covered by a diversionary carrier raid on Dutch Harbor in the Aleutians. This attack was met by a greatly outnumbered United States carrier force composed of Task Force 17 (Rear Admiral F. J. Fletcher) with Yorktown, and Task Force 16 (Rear Admiral R.A. Spruance) with Hornet and Enterprise, and by Navy, Marine Corps, and Army air units based on Midway. Planes from Midway located and attacked ships of the Japanese Occupation Force 600 miles to the west (3 June), and of the mobile Force (4 June) as it sent its aircraft against defensive installations on Midway. Concentrating on the destruction of Midway air forces and diverted by their torpedo, horizontal, and dive bombing attacks, the Japanese carriers were caught unprepared for the carrier air attack which began at 0930 with the heroic but unsuccessful effort of Torpedo Squadron 8, and were hit in full force at 1030 when dive bombers hit and sank the carriers Akagi, Kaga, and Soryu. A Japanese counter attack at noon and another 2 hours later, damaged Yorktown with bombs and torpedoes so severely that she was abandoned. In the late afternoon, U.S. carrier air hit the Mobile Force again, sinking Hiryu, the fourth and last of the Japanese carriers in action. With control of the air irretrievably lost, the Japanese retired under the attack of Midway-based aircraft (5 June) and of carrier air (6 June) in which the heavy cruiser Mikuma was sunk and the Mogami severely damaged. Japanese losses totaled two heavy and two light carriers, one heavy cruiser, 258 aircraft, and a large percentage of their experienced carrier pilots. United States losses were 40 shore-based and 92 carrier aircraft, the destroyer Hammann and the carrier the Yorktown, which sank 6 and 7 June respectively, the result of a single submarine attack. The decisive defeat administered to the Japanese put an end to their successful offensive and effectively turned the tide of the Pacific War.

4--The TBF Grumman Avenger flown by pilots of a shore-based element of Torpedo Squadron 8, began its combat career with attacks on the Japanese Fleet during the Battle of Midway.

10--Patrol planes of Pat Wing 4 discovered the presence of the enemy on Kiska and Attu--the first news of Japanese landings that had taken place on the 7th.

10--A formal organization, Project Sail, was established at NAS Quonset Point for airborne testing and associated work on Magnetic Airborne Detectors (MAD gear). This device was being developed to detect submarines by the change that they induced in the earth's magnetic field. Principal developmental efforts were being carried out by the Naval Ordnance Laboratory and the National Defense

Appendix I

The Evolution of Aircraft Carriers

Research Committee. In view of the promising results of early trials made with airships and an Army B-18, 200 sets of MAD gear were then being procured.

11-13--PBY Catalinas, operating from the seaplane tender Gillis in Nazan Bay, Atka Island, hit ships and enemy positions on Kiska in an intense 48-hour attack which exhausted the gasoline and bomb supply aboard the Gillis but was not successful in driving the Japanese from the Island.

13--Loran, long range navigation equipment, was given its first airborne test. The receiver was mounted in the K-2 airship and, in a flight from NAS Lakehurst, accurately determined position when the airship was over various identifiable objects. The test culminated with the first Loran homing from a distance 50 to 75 miles offshore during which the Loran operator, Dr. J.A. Pierce, gave instructions to the airship's commanding officer which brought them over the shoreline near Lakehurst on a course that caused the commanding officer to remark, "We weren't [just] headed for the hangar." We were headed for the middle of the hangar." The success of these tests led to immediate action to obtain operational Loran equipment.

15--Copahee, Captain J. G. Farrell commanding, was commissioned at Puget Sound Navy Yard, first of 10 escort carriers of the Bogue Class converted from Maritime Commission hulls.

16--Congress authorized an increase in the airship strength of the Navy to 200 lighter-than-air craft.

17--The development of Pelican, an antisubmarine guided missile, was undertaken by the National Defense Research Committee with Bureau of Ordnance sponsorship. This device consisted of a glide bomb which could automatically home on a radar beam reflected from the target.

17--Following the abolition of the newly created office of the Assistant Chief of Naval Operations (Air), the earlier order establishing an aviation organization in the Office of the Chief of Naval Operations was revised to the extent that the Director of the Aviation Division became responsible directly to the Vice Chief of Naval Operations.

17--A contract was awarded to Goodyear for the design and construction of a prototype model M scouting and patrol airship with 50 percent greater range and volume (625,000 cu. ft.) than the K Class. Four model M airships were procured and placed in service during World War II.

25--Preliminary investigation of early warning radar had proceeded to the point that the Coordinator for Research and Development requested development be initiated of airborne early warning radar including automatic airborne relay and associated shipboard processing and display equipment. Interest in early warning radar had arisen when Admiral King remarked to Dr. Vannevar Bush, head of the Office of Scientific Research and Development, that Navy ships need to see over the hill-i.e. beyond the line of sight.

26--Scheduled Naval Air Transport Service operations between the West coast and Alaska were initiated by VR-2.

27--The Naval Aircraft Factory was directed to participate in the development of high altitude pressure suits with particular emphasis upon testing existing types and obtaining information so that they could be tailored and fitted for use in flight. The Navy thus joined the Army which had sponsored earlier work on pressure suits. The NAF expanded its endeavors in the field of high altitude equipment which then included design of a pressure cabin airplane and construction of an altitude test chamber.

29--Following an inspection of Igor I. Sikorsky's VS-300 helicopter on 26 June, Lieutenant Commander F. A. Erickson, USCG, recommended that helicopters be obtained for antisubmarine convoy duty and life-saving.

JULY

3--In the first successful firing of an American rocket from a plane in flight, Lieutenant Commander J. H. Hean, Gunnery Officer of Transition Training Squadron, Pacific Fleet, fired a retro-rocket from a PBY-5A in flight at Goldstone Lake, Calif. The rocket, designed to be fired aft with a velocity equal to the forward velocity of the airplane, and thus to fall vertically, was designed at the California Institute of Technology. Following successful tests, the retro-rocket became a weapon complementary to the magnetic airborne detector with Patrol Squadron 63 receiving the first service installation in February 1943.

7--An agreement was reached between the Army and Navy, which provided that the Army would deliver to the Navy a specified number of B-24 Liberators, B-25 Mitchells, and B-34 Venturas to meet the Navy's requirement for long range landplanes. Also, the Navy would relinquish its production cognizance of the Boeing Renton plant to the Army for expanded B-29 production and limit its orders for PBY's to avoid interference with B-24 production.

12--Patrol Wings were reorganized to increase the mobility and flexibility of patrol aviation. Headquarters Squadrons were authorized for each wing to furnish administrative and maintenance services to attached squadrons.

Appendix I

The Evolution of Aircraft Carriers

Geographic areas of responsibility were assigned to each wing, and permanent assignment of squadrons was abolished in favor of assignment as the situation required.

19--The seaplane tender *Casco* established an advanced base in Nazan Bay, Atka, to support seaplane operations against Kiska, which included antishipping search, bombing of enemy positions, and cover for surface force bombardments.

24--The Bureau of Aeronautics issued a Planning Directive calling for procurement of four Sikorsky helicopters for study and development by Navy and Coast Guard aviation forces.

AUGUST

1--A J4F Widgeon, piloted by Ensign Henry C. White of Coast Guard Squadron 212, based at Houma, La., scored the first Coast Guard kill of an enemy submarine with the sinking of the *U-166* off the passes of the Mississippi.

7--Marine Aircraft Wings, Pacific was organized at San Diego under command of Major General Ross E. Rowell for the administrative control and logistic support of Marine Corps aviation units assigned to the Pacific Fleet. In September 1944, this command was renamed Aircraft Fleet Marine Force, Pacific.

7 August 1942--9 February 1943--Capture of Guadalcanal--Air support for the U.S Marines' first amphibious landing of World War II was provided by three carriers of Air Support Force (Rear Admiral L. Noyes), and by Navy, Marine, and Army units of Aircraft, South Pacific (Rear Admiral J. S. McCain) operating from bases on New Caledonia and in the New Hebrides. Carrier forces withdrew from direct support (9 Aug) but remained in the area to give overall support to the campaign during which they participated in several of the naval engagements fought over the island. *Saratoga* sank the Japanese light carrier Ryujo in the Battle of the Eastern Solomons (23-25 Aug); *Enterprise* was hit by carrier-based bombers (24 Aug) and forced to retire; *Saratoga* was damaged by a submarine torpedo (31 Aug) and forced to retire; and Wasp was sunk by a submarine (15 Sep) while escorting a troop convoy to Guadalcanal. *Hornet*, in Task Group 17 (Rear Admiral G. D. Murray), hit targets in the Buin-Tonolei-Faisi area (5 Oct); attacked beached Japanese transports and supply dumps on Guadalcanal; destroyed a concentration of seaplanes at Rekata Bay (16 Oct); and, with the Enterprise, fought in the Battle of Santa Cruz (26-27 Oct) in which she was sunk by air attack. In final carrier actions of the campaign, the Enterprise took part in the last stages of the Naval Battle for Guadalcanal (12-15 Nov), assisting in sinking 89,000 tons of war and cargo ships, and in the Battle

Appendix I

The Evolution of Aircraft Carriers

of Rennel Island (29-30 Jan) in which two escort carriers also participated. Ashore, air forces in great variety provided direct support. Navy patrol squadrons flew search, rescue, and offensive missions from sheltered coves and harbors. Marine Fighter Squadron 223 and Scout Bombing Squadron 232, delivered by the escort carrier Long Island, initiated operations from Henderson Field on Guadalcanal (20 Aug) and were joined within a week by AAF fighter elements and dive bombers from the *Enterprise*, and by other elements as the campaign progressed. Until the island was secure (9 Feb), these forces flew interceptor patrols, offensive missions against shipping, and close air support for the Marines and for Army troops relieving them (13 Oct). Marine air units carrying the major air support burden accounted for 427 enemy aircraft during the campaign.

10--The headquarters of Patrol Wing 3 shifted within the Canal Zone from NAS Coco Solo to Albrook Field for closer coordination with the Army Air Force Command in the defense of the Panama Canal.

12--*Cleveland* (CL 55), operating in the Chesapeake Bay, demonstrated effectiveness of the radio-proximity fuse against aircraft by destroying three radio-controlled drones with four proximity bursts fired from her 5-inch guns. This successful demonstration led to mass production of the fuse.

12--*Wolverine* (IX 64) was commissioned at Buffalo, Commander G.R. Fairlamb commanding. This ship and *Sable* (IX 81) commissioned the following May, were Great Lakes excursion ships converted for aviation training and as such they operated for the remainder of the war on the inland waters of Lake Michigan. They provided flight decks upon which hundreds of student naval aviators qualified for carrier landings and many flight deck crews received their first practical experience in handling aircraft aboard ship.

13--The Commander in Chief U.S. Fleet directed that an Aircraft Experimental and Developmental Squadron be established about 30 September 1942 at NAS Anacostia. This squadron, which replaced the Fleet Air Tactical Unit, was to conduct experiments with new aircraft and equipment in order to determine their practical application and tactical employment.

15--Patrol Wing 11 was established at Norfolk, Commander S.J. Michael commanding. Five days later the Wing moved to San Juan, P.R., for operations under the Caribbean Sea Frontier.

20--The designation of escort carriers was changed from AVG to ACV.

Appendix I

The Evolution of Aircraft Carriers

24--*Santee*, Captain W. D. Sample commanding, was placed in commission at the Norfolk Navy Yard; the first of four escort carriers of the Sangamon Class converted from Cimarron Class fleet oilers.

30--The occupation of Adak by Army forces and the establishment of an advanced seaplane base there by the tender Teal, put North Pacific forces within 250 miles of occupied Kiska and in a position to maintain a close watch over enemy shipping lanes to that island and to Attu. The tender *Casco*, conducting support operations from Nazan Bay, was damaged by a submarine torpedo and temporarily beached.

SEPTEMBER

1--U.S. Naval Air Forces, Pacific, Rear Admiral A. W. Fitch commanding, was established for the administrative control of all air and air service units under the Commander in Chief, Pacific, replacing the offices of Commander Carriers, Pacific, and Commander Patrol Wings, Pacific. The subordinate commands Fleet Air West Coast, Fleet Air Seattle, and Fleet Air Alameda were established at the same time.

6--The first Naval Air Transport Service flight to Argentina, Newfoundland, marked the beginning of air transport expansion along the eastern seaboard that during the month extended briefly to Iceland and reached southward to the Canal Zone and Rio de Janeiro.

7--Air Transport Squadron 2, based at Alameda, established a detachment at Pearl Harbor and began a survey flight to the South Pacific as a preliminary to establishing routes between San Francisco and Brisbane, Australia.

16--Patrol Wing 12 was established at Key West, Captain W. G. Tomlinson commanding, for operations under the Gulf Sea Frontier.

19--Commander Patrol Wing 1 departed Kaneohe, Hawaii, for the South Pacific to direct the operations of patrol squadrons already in the area. Headquarters were first established at Noumea, New Caledonia, and subsequently at Espiritu Santo, Guadalcanal, and Munda.

OCTOBER

1--Airship Patrol Group 3, Captain Scott E. Peck commanding, was established at Moffett Field to serve as the administrative command for airship squadrons operating on the west coast.

Appendix I

The Evolution of Aircraft Carriers

1--Three functional training commands were established for Air Technical Training, Air Primary Training, and Air Intermediate Training, with headquarters initially at Chicago, Kansas City, and Pensacola respectively.

12--Naval Air Centers Hampton Roads, San Diego, Seattle, and Hawaiian Islands, and Naval Air Training Centers Pensacola and Corpus Christi, were established to consolidate under single commands the complex of naval aviation facilities that had become operational in the vicinity of certain large air stations.

15--Patrol Wing 14, Captain W. M. McDade commanding, was established at San Diego for operations under the Western Sea Frontier and for duties concerned with equipping, forming, and establishing patrol squadrons.

17--Inshore Patrol Squadrons (VS), engaged in coastal antisubmarine reconnaissance and convoy duty under the Sea Frontiers, were transferred to Patrol Wings for administrative control.

19--The initial installation and deployment of the ASB-3 airborne search radar was reported. This radar, developed by the Naval Research Laboratory for carrier based aircraft, had been installed in five TBF-1's by NAS New York and five SBD-3's by NAS San Pedro. One aircraft of each type was assigned to Air Group Eleven (Saratoga) and the others shipped to Pearl Harbor. The remaining sets on the initial contract for 25 were assigned as spare parts and for purposes of training.

22--Westinghouse Electric and Manufacturing Company, by amendment to a design study contract, was authorized to construct two 19A axial flow turbojet power plants. Thereby, fabrication was initiated of the first jet engine of wholly American design.

28--Procurement of the expendable radio sonobuoy for use in antisubmarine warfare was initiated as the Commander in Chief, U.S. Fleet directed the Bureau of Ships to procure 1,000 sonobuoys and 100 associated receivers.

31--Air Transport Squadrons Pacific was established over the NATS squadrons based in the Pacific and those on the west coast flying the mainland to Hawaii routes.

NOVEMBER

1--Patrol Wings were redesignated Fleet Air Wings, and to permit the organization of patrol aviation on the task force principle, the practice of assigning a standard number of squadrons to each Wing was changed to provide

Appendix I

The Evolution of Aircraft Carriers

for the assignment of any and all types of aircraft required by the Wing to perform its mission in its particular area.

1--Airship Patrol Group 1 at NAS Lakehurst was redesignated Fleet Airship Group 1.

2--NAS Patuxent River was established to serve as a facility for testing experimental airplanes, equipment and material, and as a NATS base.

2--Fleet Air Wing 6, Captain D. P. Johnson commanding was established at NAS Seattle.

8-11--Invasion of North Africa--Carrier aircraft from *Ranger* and the escort carriers *Sangamon*, *Suwannee*, and *Santee* of Task Group 34.2 (Rear Admiral E.D. McWhorter) of the Western Naval Task Force, covered the landings of Army troops near Casablanca (8 Nov) and supported their operation ashore until opposing French forces capitulated (11 Nov). The escort carrier *Chenango* accompanied assault forces to the area and launched her load of 78 AAF P-40s (10-11 Nov) for operations from the field at Port Lyautey.

13--Patrol Squadron 73 arrived at Port Lyautey from Iceland via Bally Kelly, Ireland, and Lyncham, England. Supported by the seaplane tender *Barnegat*, the squadron began antisubmarine operations from French Morocco over the western Mediterranean, the Strait of Gibraltar, and its approaches. Patrol Squadron 92 also arrived at Port Lyautey on the same day via Cuba, Brazil, Ascension Island, and West Africa.

16--Naval aviation's first night fighter squadron, VMF(N)-531, was established at MCAS Cherry Point with Lieutenant Colonel Frank H. Schwable in command. After initial training with SNJs and SB2A-4s, the squadron was assigned twin-engined PV-1 aircraft equipped with British Mark IV type radar.

23--The V-173, a full-scale model of a fighter aircraft with an almost circular wing, made its first flight at the Vought-Sikorsky plant, Stratford, Conn. A military version of this aircraft, the XF5U-1, was constructed later but never flown.

DECEMBER

1--Fleet Air Wing 15, Captain G. A. Seitz commanding, was established at Norfolk for operations under the Moroccan Sea Frontier.

Appendix I

The Evolution of Aircraft Carriers

1--Fleet Airship Wing 30, Captain George H. Mills commanding, was established at NAS Lakehurst to administer Atlantic Fleet Airship Groups and their component squadrons.

1--Airship Patrol Group 3 at NAS Moffett Field was redesignated Fleet Airship Wing 31.

26--The Chief of Naval Operations approved the merger of the Service Force Aviation Repair Unit and Advanced Cruiser Aircraft Training Unit, established in October 1941 and June 1942 respectively, to form a Scout Observation Service Unit (SOSU) with a mission to maintain battleship and cruiser aircraft and to indoctrinate pilots in their specialized operations. This SOSU, the first of three established during World War II was established 1 January 1943.

27--*Santee*, first of 11 escort carriers assigned to Hunter-Killer duty, sortied Norfolk with Air Group 29 on board for free-roving antisubmarine and anti-raider operations in the South Atlantic.

31--After pointing out that the need for airborne radar was so apparent and urgent that peacetime methods of procurement and fleet introduction could not be followed, the Chief of the Bureau of Aeronautics requested the Naval Research Laboratory to continue to provide personnel capable of assisting fleet units in the operation and maintenance of radar equipment until a special group of trained personnel could be assembled for that purpose. This special group developed within a few months into the Airborne Coordination Group which provided trained civilian electronics specialists to fleet units throughout the war and into the post-war period.

31--*Essex*, Captain D. B. Duncan commanding, was placed in operating status at Norfolk; the first of 17 ships of her class commissioned during World War II.

1943

JANUARY

1--Naval Reserve Aviation Bases (NRAB) engaged in Primary Flight Training in all parts of the country were redesignated Naval Air Stations (NAS) without change of mission. This was the end of the NRAB's except for Anacostia, which was abolished on 7 July 1943, and Squantum which became an NAS on 1 September 1943.

1--Air Force, Atlantic Fleet, was established, Rear Admiral A. D. Bernhard commanding, to provide administrative, material, and logistic services for

Appendix I

The Evolution of Aircraft Carriers

Atlantic Fleet aviation in place of the former separate commands Fleet Air Wings, Atlantic, and Carriers, Atlantic, which were abolished. By the same order Fleet Air, Quonset, was established as a subordinate command.

1--Ground Controlled Approach equipment (GCA) was called into emergency use for the first time when a snowstorm closed down the field at NAS Quonset Point a half hour before a flight of PBYs was due to arrive. The GCA crew located the incoming aircraft on their search radar, and using the control tower as a relay station, "talked" one of them into position for a contact landing. This recovery was made only 9 days after the first successful experimental demonstration of GCA.

5--The first combat use of a proximity fuzed projectile occurred when *Helena* (CL 50) off the south coast of Guadalcanal, destroyed an attacking Japanese dive bomber with the second salvo from her 5-inch guns.

7--A change in the pilot training program was implemented by the opening of Flight Preparatory Schools in 20 colleges and universities in all parts of the country. Under the new program, students began their training at these schools with 3 months of academic work fundamental to ground school subjects, then proceeded to War Training Service courses conducted by the Civil Aeronautics Administration at universities for 2 months's training in ground subjects and elementary flight under civilian instructors; then to the Pre-Flight Schools for 3 months of physical conditioning; and finally to Navy flight training beginning at one of the Primary Training Bases.

7--Development of the first naval aircraft to be equipped with a turbojet engine was initiated with the issuance of a Letter of Intent to McDonnell Aircraft Corporation for engineering, development, and tooling for two VF airplanes. Two Westinghouse 19-B turbojet engines were later specified and the aircraft was designated XFD-1. It became the prototype for the FH-1 Phantom jet fighter.

10--Fleet Air Wing 15 headquarters was transferred from Norfolk to Port Lyautey, French Morocco, to direct patrol plane operations in the Mediterranean and Gibraltar Strait area.

12--The Chief of Naval Air Operational Training directed that aircraft operating from stations under his command be marked for identification purposes with letters and numerals in three groups separated by a dash. The first group provided a letter identification of the station, the second a letter identifying the unit type and the third the number of the aircraft in the unit. The order also provided that when more than one unit was on board a station, a number be

Appendix I

The Evolution of Aircraft Carriers

added to the station letter. Thus J2-F-22 identified the aircraft as from Jacksonville, OTU #2 Fighter Training Unit, plane number 22.

14--*Independence*, Captain G.R. Fairlamb Jr., commanding, was placed in commission at Philadelphia; the first of nine light carriers of her class constructed on *Cleveland* Class cruiser hulls.

15--Captain H.S. "Seth" Warner, Head of the Flight Statistics Desk of the Bureau of Aeronautics, introduced Grampa Pettibone, in the BuAer News Letter. Pettibone, a cartoon character drawn by Lieutenant Robert Osborn, was produced as a safety feature in the hope of cutting down on pilot-error accidents. Gramps went on to become famous through the post-war decades as Osborn, after leaving the Navy, continued to contribute his character to Naval Aviation News magazine.

17--Following tests conducted at NAS San Diego by six experienced pilots flying F4U-1s, the Commanding Officer of VF-12, Commander J.C. Clifton, reported that anti-blackout suits raised their tolerance to accelerations encountered in gunnery run and other maneuvers by three to four Gs.

FEBRUARY

1--Bombing Squadron, VB-127, was established at NAS Deland, Fla., with Lieutenant Commander William K. Gentner in command. The squadron was equipped with PV-1 Venturas and, although not the first land plane patrol squadron in the Navy, was the first to have the VB designation.

1--A new specification prescribing color and marking of naval aircraft became effective. A basic camouflage color scheme was provided for use on fleet aircraft which consisted of semi-gloss sea blue on surfaces viewed from above and non-specular insignia white on surfaces viewed from below. The terminology "basic non-camouflage" and "maximum visibility" were introduced for the color schemes described in April 1942, and used on intermediate and primary trainers.

1--Regulations governing display of National Insignia on aircraft were again revised by the order to remove those on the upper right and lower left wing surfaces.

11--A contract was issued to the Ryan Aeronautical Corporation for the XFR-1 fighter. This aircraft incorporated a conventional reciprocating engine for use in normal operations and the turbojet for use as a booster during takeoffs and maximum performance flights. Development and production were handled on a

Appendix I
The Evolution of Aircraft Carriers

crash basis to equip escort carrier squadrons at the earliest possible date. However, numerous bugs were encountered which prevented the FR-1's assignment to combat.

11--The Vought F4U Corsair was flown on a combat mission for the first time when 12 planes of VMF-124 based on Guadalcanal escorted a PB2Y Dumbo to Vella Lavella to pick up downed pilots. The flight was uneventful. Its first combat action came 2 days later when pilots from the same squadron ran into air opposition while escorting PB4Ys of VP-51 on a daylight strike against enemy shipping in the Kahili area of Bougainville.

13--The Naval Air Transport Service was reorganized and the establishment of Wings was directed for the Atlantic and west coast squadrons.

15--The Commander in Chief U.S. Fleet assigned responsibility for sea-going development of helicopters and their operation in convoys to the Coast Guard and directed that tests be carried out to determine if helicopters operating from merchant ships would be of value in combating submarines.

16--Fleet Air Wing 16, Captain R. D. Lyon commanding, was established at Norfolk.

17--Lighter-than-air operations over the Caribbean were initiated from Edinburgh Field, Trinidad, by the K-17 of Airship Patrol Squadron 51.

19--A letter of intent was issued to Vega Airplane Company for two XP2V-1 patrol planes, thereby initiating development of the P2V Neptune series of land-based patrol aircraft.

21 February--1 November--Advance up the Solomons Chain--In a series of amphibious operations, directly and indirectly supported by Marine Corps, Navy and Army units of Aircraft, South Pacific, and Aircraft, Solomons, Central Pacific Forces moved from Guadalcanal up the Solomon Islands towards the Japanese naval base at Rabaul. Beginning with the unopposed landing in the Russells (21 Feb), these forces leapfrogged through the islands establishing bases and airfields as they went. Moving into Segi of the New Georgia Group (21 June), through Rendova, Onaivisi, Wickham Anchorage, Kiriwini and Woodlark (30 June), Viru (2 July), Zanana (2 July), Rice Anchorage (5 July), Vella Lavella (15 Aug), Arundel (27 Aug), and Treasury Islands (27 Oct), they reached Bougainville where landings on Cape Torokina were additionally supported by carrier air strikes (1, 2 Nov) on the Buka-Bonis airfields.

Appendix I

The Evolution of Aircraft Carriers

24--The Naval Photographic Science Laboratory was established at NAS Anacostia under the direction of the Bureau of Aeronautics to provide photographic services to the Navy and to develop equipment and techniques suitable for fleet use.

MARCH

1--Air Transport Squadrons, West Coast, was established at NAAS Oakland with control over all NATS squadrons west of the Mississippi except those on the mainland to Honolulu run.

1--A revision of the squadron designation system changed Inshore Patrol Squadrons to Scouting Squadrons (VS), Escort Fighting Squadrons (VGF) to Fighting Squadrons (VF), Escort Scouting Squadrons (VGS) to Composite Squadrons (VC) and Patrol Squadrons (VP) operating land type aircraft to Bombing Squadrons (VB). This revision also redesignated carrier Scouting Squadrons (VS) as VB and VC and as a result the types of squadrons on Essex Class carriers was reduced to three. In spite of this change, the aircraft complement of their Air Groups remained at its previous level of 21 VF, 36 VSB and 18 VTB.

1--Fleet Airship Group 2, Captain W. E. Zimmerman commanding, was established at NAS Richmond, Florida, and placed in charge of lighter-than-air operations in the Gulf Sea Frontier.

4--Changes to the characteristics of Essex Class carriers were authorized by the Secretary, including installation of a Combat Information Center (CIC) and Fighter Director Station, additional anti-aircraft batteries, and a second flight deck catapult in lieu of one athwartships on the hangar deck.

5--Bogue, with VC-9 on board, joined Task Group 24.4 at Argentia and began the escort of convoys to mid-ocean and return. Although the Santee had previously operated on Hunter-Killer duty, Bogue was the center of the first of the Hunter-Killer groups assigned to convoy escort.

15--Fleet Air Wing 4 headquarters moved westward on the Aleutian chain from Kodiak to Adak.

20--Forty-two Navy and Marine Corps Avengers, on a night flight from Henderson Field, mined Kahili Harbor, Bougainville. A coordinated attack on Kahili airfield by AAF heavy bombers contributed to the success of this, the first aerial mining mission in the South Pacific.

Appendix I

The Evolution of Aircraft Carriers

23--The Training Task Force Command was established with headquarters at NAS Clinton, Okla., to form, outfit and train special units for the operational employment of assault drone aircraft.

29--Tests of forward firing rockets projectiles from naval aircraft were completed at the Naval Proving Ground, Dahlgren, using an SB2A-4 aircraft.

29--Air Transport Squadrons, Atlantic, was commissioned at Norfolk to supervise and direct operations of NATS squadrons based on the Atlantic seaboard.

APRIL

1-Aircraft Antisubmarine Development Detachment, Commander A.B. Vosseller in command, was established at NAS Quonset Point under Air Force, Atlantic Fleet, to develop tactical training programs and techniques that would make full use of newly developed countermeasures equipment.

1--The first Navy night fighter squadron, VF(N)-75, was established at Quonset Point, Commander W. J. Widhelm, commanding.

4--The Naval Aircraft Factory reported that, in tests of an automatic flying device for use on towed gliders, the LNT-1 had been towed automatically without assistance from the safety pilot.

14--Fleet Air Wing 16 transferred from Norfolk to Natal, Brazil, to direct patrol plane antisubmarine operations under the Fourth Fleet in the South Atlantic.

21--Captain Frederick M. Trapnell made a flight in the Bell XP-59A jet Airacomet at Muroc, Calif., the first jet flight by a U.S. Naval Aviator.

MAY

3--Air Transport Squadron 1 (VR-1), based at Norfolk, extended the area of its operations with a flight to Prestwick, Scotland, via Reykjavik, Iceland. This was the first R5D operation in the Naval Air Transport Service.

4--The first regular patrols began from Amchitka, extending the search coverage by Fleet Air Wing 4 beyond Attu toward the Kuriles.

4--To expedite the evaluation of the helicopter in antisubmarine operations, the Commander in Chief, U.S. Fleet directed that a "joint board" be formed with

Appendix I
The Evolution of Aircraft Carriers

representatives of the Commander in Chief, U.S. Fleet; the Bureau of Aeronautics; the Coast Guard; the British Admiralty and the Royal Air Forces. The resulting Combined Board for the Evaluation of the Ship-Based Helicopter in Anti-Submarine Warfare was later expanded to include representatives of the Army Air Forces, the War Shipping Administration and the National Advisory Committee for Aeronautics.

7--Navy representatives witnessed landing trials of the XR-4 helicopter aboard the merchant tanker Bunker Hill in a demonstration sponsored by the Maritime Commission and conducted in Long Island Sound. The pilot, Colonel R. F. Gregory, AAF, made about 15 flights, and in some of these flights he landed on the water before returning to the platform on the deck of the ship.

11-30--Occupation of Attu--Air support for the landing of Army troops (11 May) and for their operations ashore was provided by Navy and Marine units on the escort carrier Nassau (11-20 May), and by the Navy and Army units of North Pacific Force (11-20 May). This was the first use of CVE based aircraft in air support in the Pacific and the debut of a Support Air Commander afloat. His team consisted of three officers and a radioman and his post was a card table aboard Pennsylvania. Colonel W. O. Eareckson, USA, an experienced Aleutian pilot, was in command of the unit.

15--The Naval Airship Training Command was established at Lakehurst to administer and direct lighter-than-air training programs at the Naval Air Centers, Lakehurst and Moffett Field, and to direct the Experimental and Flight Test Department at Lakehurst.

18--The program for the use of gliders as transports for Marine Corps combat troops was canceled, thereby returning the Navy's glider development to an experimental basis.

22--Grumman Avengers of VC-9, based on Bogue, attacked and sank the submarine U-569 in the middle north Atlantic scoring the first sinking of the war by escort carriers on hunter-killer patrol.

24--Special Project Unit Cast was organized at NAS Squantum to provide, under Bureau of Aeronautics direction, the services required to flight test the electronics equipment being developed at the Radiation and Radio Research Laboratories.

Appendix I

The Evolution of Aircraft Carriers

JUNE

7--The establishment of NAF Attu, within 1 week of its capture from the Japanese, brought Fleet Air Wing 4 bases to the tip of the Aleutian chain, nearly 1,000 miles from the Alaskan mainland and 750 miles from Japanese territory in the Kuriles.

7--The Commander in Chief, U.S. Fleet established a project for airborne test, by Commander Fleet Air, West Coast, of high velocity, "forward shooting" rockets. These rockets, which had nearly double the velocity of those tested earlier at Dahlgren, had been developed by a rocket section, led by Dr. C.C. Lauritsen, at the California Institute of Technology under National Defense Research Committee auspices and with Navy support. This test project, which was established in part on the basis of reports of effectiveness in service of a similar British rocket, completed its first airborne firing from a TBF of a British rocket on 14 July and of the CalTech round on 20 August. The results of these tests were so favorable that operational squadrons in both the Atlantic and Pacific Fleets were equipped with forward firing rockets before the end of the year.

10--Lieutenant Commander F. A. Erickson, USCG, proposed that helicopters be developed for antisubmarine warfare, "not as a killer craft but as the eyes and ears of the convoy escorts." To this end he recommended that helicopters be equipped with radar and dunking sonar.

15--President Roosevelt approved a ceiling of 31,447 useful planes for the Navy.

28--A change in the design of the National Star Insignia added white rectangles on the left and right sides of the blue circular field to form a horizontal bar, and a red border stripe around the entire design. The following September, Insignia Blue was substituted for the red.

29--NAS Patuxent River began functioning as an aircraft test organization with the arrival of the Flight Test unit from NAS Anacostia.

29--Elements of VP-101 arrived at Brisbane from Perth, thereby extending the patrol coverage of Fleet Air Wing 10 to the east coast of Australia and marking the beginning of a northward advance of patrol operations toward the Papuan Peninsula of New Guinea.

Appendix I

The Evolution of Aircraft Carriers

JULY

5--The first turbojet engine developed for the Navy, the Westinghouse 19A, completed its 100-hour endurance test.

8--Casablanca, first of her class and first escort carrier designed and built as such, was placed in commission at Astoria, Oreg., Captain S. W. Callaway commanding.

14--The Secretary issued a General Order forming the Naval Air Material Center, consisting of the separate commands of the Naval Aircraft Factory, the Naval Aircraft Modification Unit, the Naval Air Experimental Station and the Naval Auxiliary Air Station. This action, effective 20 July, consolidated in distinct activities the production, modification, experimental, and air station facilities of the former Naval Aircraft Factory organization.

15--New designations for carriers were established which limited the previous broadly applied CV symbol to Saratoga, Enterprise and carriers of the Essex Class, and added CVB (Aircraft Carriers, Large) for the 45,000 ton class being built and CVL (Aircraft Carriers, Small) for the 10,000 ton class built on light cruiser hulls. The same directive reclassified escort carriers as combatant ships and changed their symbol from ACV to CVE.

15--The airship organization of the U.S. Fleet was modified. Fleet Airship Wings 30 and 31 were redesignated Fleet Airships, Atlantic, and Pacific respectively. Airship Patrol Groups became Airship Wings. Airship Patrol Squadrons became Blimp Squadrons, and the addition of two more wings and the establishment of Blimp Headquarters Squadrons in each wing was authorized.

18--The airship K-74, while on night patrol off the Florida coast, attacked a surfaced U-boat and in the gun duel which followed was hit and brought down--the only airship lost to enemy action in World War II. The submarine, U-134, was damaged enough to force her return to base, and after surviving two other attacks on the way, was finally sunk by British bombers in the Bay of Biscay.

19--The Naval Aircraft Factory was authorized to develop the Gorgon, an aerial ram or air-to-air missile powered by a turbojet engine and equipped with radio controls and a homing device. The Gorgon was later expanded into a broad program embracing turbojet, ramjet, pulsejet, and rocket power; straight wing, swept wing, and canard (tail first) air frames; and visual, television, heat-homing, and three types of radar guidance for use as air-to-air, air-to-surface and surface-to-surface guided missiles and as target drones.

Appendix I

The Evolution of Aircraft Carriers

22--Since there had been no operational need for arresting gear and related equipment for landing over the bow of aircraft carriers, the Vice Chief of Naval Operations approved its removal.

23--Patrol Squadron 63, the first U.S. Navy squadron to operate from Great Britain in World War II, arrived at Pembroke Dock, England, to assist in the antisubmarine patrol over the Bay of Biscay.

AUGUST

2--Fleet Airship Wings 4 and 5, Captain W. E. Zimmerman and Commander John D. Reppy commanding, were established at Maceio, Brazil, and Edinburgh Field, Trinidad, for anti-submarine and convoy patrols in the South Atlantic and southern approaches to the Caribbean.

4--The Chief of Naval Air Intermediate Training directed that Aviation Safety Boards be established at each training center under his command.

5--CominCh directed the use of Fleet Air Wing commanders in subordinate commands of Sea Frontiers and suggested their assignment as Deputy Chiefs of Staff for Air.

15--The arrival of Aircraft Experimental and Development Squadron (later Tactical Test) from NAS Anacostia to NAS Patuxent River completed the transfer of aircraft test activities.

15--The landing of U.S. Army and Canadian troops on Kiska by a Naval Task Force made the first use in the Pacific of Air Liaison Parties (ALP) with forces ashore. Although the enemy had deserted the island, the landing provided opportunity to prove that the principle of the ALP was sound and that rapid and reliable voice communications between front line commanders and the Support Air Control Unit afloat were possible.

18--To give naval aviation authority commensurate with its World War II responsibility, the Secretary of the Navy established the Office of the Deputy Chief of Naval Operations (Air), charging it with responsibility for "the preparation, readiness and logistic support of the naval aeronautic operating forces. By other orders issued the same day, five divisions were transferred from the Bureau of Aeronautics to form the nucleus of the new office and Vice Admiral J. S. McCain took command as the first DCNO (Air).

Appendix I

The Evolution of Aircraft Carriers

21--Headquarters of Fleet Air Wing 7 was established at Plymouth, England, to direct patrol plane operations against submarines in the Bay of Biscay, the English Channel and the southwest approaches to England.

29--The formation of combat units for the employment of assault drone aircraft began within the Training Task Force Command as the first of three Special Task Air Groups was established. The component squadrons, designated VK, began establishing on 23 October.

31--Second Strike on Marcus--Task Force 15 (Rear Admiral C. A. Pownall), built around *Essex*, the new *Yorktown* and *Independence* launched nine strike groups in a day-long attack on Japanese installations on Marcus Island, the first strikes by *Essex* and *Independence* Class carriers, and the first combat use of the F6F Grumman Hellcat.

SEPTEMBER

1--Two light carriers of Task Group 11.2 (Rear Admiral A. W. Radford) and Navy patrol bombers from Canton Island furnished day and night air cover for naval units landing occupation forces on Baker Island, east of the Gilberts.

15--Fleet Air Wing 17, Commodore T. S. Combs commanding, was established at Brisbane, Australia, for operations in the Southwest Pacific area.

15--French Patrol Squadron 1 (VFP-1), manned by "Fighting French" naval personnel trained under U. S. Navy control, was established at NAS Norfolk.

18--A three-carrier task force (Rear Admiral C. A. Pownall), attacked Tarawa, Makin, and Abe Mama Atolls in the Gilbert Islands.

18--Training was assigned as a primary mission to Fleet Air Wing 5 at Norfolk, and Fleet Air Wing 9 assumed responsibility for all patrol plane operations in the Eastern Sea Frontier.

27--The beginning of airship operations in the South Atlantic was marked by the arrival of the K-84, of Blimp Squadron 41, at Fortaleza, Brazil.

30--An advance detachment of Bombing Squadron 107, equipped with PB4Y Liberators, arrived at Ascension Island to join AAF units on antisubmarine barriers and sweeps across the narrows of the South Atlantic.

Appendix I

The Evolution of Aircraft Carriers

OCTOBER

1--Air Force, Atlantic Fleet, was reorganized and Fleet Air, Norfolk, and Fleet Airships, Atlantic, were established as additional subordinate commands.

1--The authorized complement of fighters in Essex Class Carrier Air Groups was raised, increasing the total aircraft normally on board to 36 VF, 36 VB and 18 VT. The authorized complement for CVL groups was established at the same time as 12 VF, nine VB and nine VT and revised in November 1943 to 24 VF and nine VT and remained at that level through the war.

4--In conjunction with her duties in protecting North Atlantic convoy routes to Russia, Ranger launched two strikes against German shipping in Norway--one in and around Bodo Harbor; the other along the coast from Alter Fjord to Kunna Head.

5--Coast Guard Patrol Squadron 6 was established at Argentia, Newfoundland, Commander D. B. MacDiarmid, USCG, commanding, to take over the rescue duties being performed by naval aircraft in Greenland and Labrador.

5-6--Second Wake Raid--Task Force 14 (Rear Admiral A. E. Montgomery), composed of six new carriers, seven cruisers, and 24 destroyers, making it the largest carrier task force yet assembled, bombed and bombarded Japanese installations on Wake Island. In the course of the 2-day strikes, ship handling techniques for a multicarrier force, devised by Rear Admiral F. C. Sherman's staff on the basis of experience in the South Pacific, were tested under combat conditions. Lessons learned from operating the carriers as a single group of six, as two groups of three, and as three groups of two, provided the basis for many tactics which later characterized carrier task force operations.

6--The Naval Airship Training Command at Lakehurst was redesignated the Naval Airship Training and Experimental Command.

12--The Bureau of Ordnance established a production program for 3,000 Pelican guided missiles at a delivery rate of 300 a month.

16--The Navy accepted its first helicopter, a Sikorsky YR-4B (HNS-1), at Bridgeport, Connecticut, following a 60 minute acceptance test flight by Lieutenant Commander F. A. Erickson, USCG.

31--Lieutenant H. D. O'Neil of VF(N)-75, operating from Munda, New Georgia, destroyed a Betty during a night attack off Vella Lavella, the first kill by a radar-equipped night fighter of the Pacific Fleet. Major T. E. Hicks and Tech Sergeant Gleason from VMF(N)-531 provided ground-based fighter direction.

NOVEMBER

1--A detachment of Bombing Squadron 145, equipped with Venturas, began operations from Fernando Noronha Island, extending the area of Fleet Air Wing 16 antisubmarine patrols over the South Atlantic toward Ascension Island.

5--First Rabaul Strike--A two-carrier task force (Rear Admiral F. C. Sherman) delivered an air attack on the naval base at Rabaul damaging several warships of the Japanese Second Fleet.

8--The Chief of Naval Operations directed that Aviation Safety Boards, similar to those in the Intermediate Training command, be established in the Primary and Operational Training Commands.

8--The Naval Ordnance Test Station, Inyokern, California, was established for research, development and testing weapons and to provide primary training in their use. It initially supported the California Institute of Technology which, through the Office of Scientific Research and Development, was undertaking the development and testing of rockets, propellants and launchers.

11--Second Rabaul Strike--Three heavy and two light carriers organized in two carrier task forces (Rear Admirals F. C. Sherman and A. E. Montgomery), hit Japanese naval shipping at Rabaul sinking one destroyer and damaging ships, including two cruisers. In this attack SB2C Curtiss Helldivers were used in combat for the first time.

13-19--Army and Navy aircraft of Task Force 57 (Rear Admiral J. H. Hoover), based on islands of the Ellice, Phoenix, and Samoan Groups and on Baker Island, conducted long-range night bombing attacks on Japanese bases in the Gilbert and Marshall Islands as a preliminary to the invasion of the Gilberts.

18-26--Occupation of the Gilbert Islands--Six heavy and five light carriers of Task Force 50 (Rear Admiral C. A. Pownall) opened the campaign to capture the Gilberts with a 2-day air attack on airfields and defensive installations in the islands (18-19 Nov), covered the landings of Marines and Army troops on Tarawa and Makin Atolls (20 Nov) and on Abemama (21 Nov), and supported their operations ashore (21-24 Nov). Eight escort carriers, operating with the Attack Forces, covered the approach of assault shipping (10-18 Nov), flew antisubmarine and combat air patrols in the area, and close support missions on call (19-24 Nov). After the islands were secure (24 Nov), one carrier group

Appendix I

The Evolution of Aircraft Carriers

remained in the area for another week as a protective measure. The first unit of the garrison air force, VF-1, took off from the escort carriers Barnes and Nassau (25 Nov) and landed on Tarawa airstrip. One escort carrier, Liscome Bay was lost (24 Nov) to submarine attack, and the light carrier Independence was damaged (20 Nov) by air attack. The first attempts at night interception from carriers were made during the campaign by a team of two Hellcats and one radar equipped Avenger operating from Enterprise and led by the Air Group Commander, Lieutenant Commander E. H. (Butch) O'Hare. In operation the fighters flew wing on the Avenger and after being vectored to the vicinity of the enemy aircraft by the ship's fighter director relied on the Avenger's radar to get within visual range. On the first occasion (24 Nov) no intercepts were made but on the second (26 Nov) the enemy was engaged in the first aerial battle of its type which so disrupted the attack that the flight was credited with saving the task group from damage.

27--The first of the Martin Mars flying boats was delivered to VR-8 at NAS Patuxent River, Md.

30--On her first operational assignment, the Martin Mars, in the hands of Lieutenant Commander W. E. Coney and crew of 16, took off from Patuxent River carrying 13,000 pounds of cargo that was delivered at Natal, Brazil, in a nonstop flight of 4,375 miles and of 28 hours 25 minutes duration.

30--A department of Aviation Medicine and Physiological Research was authorized at the Naval Air Material Center, to study physiological factors particularly as related to design of high speed and high altitude aircraft.

DECEMBER

1--Aircraft, Central Pacific, Rear Admiral J.H. Hoover commanding, was established under Commander, Central Pacific, for operational control of defense forces and shore-based air forces in the area.

1--The Naval Air Ferry Command was established as a Wing of the Naval Air Transport Service. It assumed the functions previously performed by Aircraft Delivery Units in ferrying new aircraft from contractor plants and modification centers to embarkation points for ultimate delivery to the Fleet.

4--At the close of the Gilberts Campaign, two groups of Task Force 50 (Rear Admiral C. A. Pownall) composed of four heavy and two light carriers and screening ships, bombed airfields and shipping at Wotje and Kwajalein Atolls in the Marshall Islands.

Appendix I

The Evolution of Aircraft Carriers

8--A striking force of two carriers, six battleships, and 12 destroyers bombed and bombarded enemy installations on Nauru, to the west of the Gilberts.

15--Observation Fighter Squadron 1 (VOF-1), first of three of its type brought into existence during World War II, was established at Atlantic City with Lieutenant Commander W. F. Bringle in command.

17--Commander Aircraft, Solomons, joined in the air campaign to reduce the Japanese Naval Base at Rabaul with a fighter sweep of Navy, Marine Corps, and New Zealand planes led by Marine Ace Major Gregory Boyington. Intensive follow-up attacks through February 1944 assisted in the establishment of encircling allied bases. Rabaul remained under air attack until the war's end, the last strike being delivered by Marine Corps PBJ's on 9 August 1945.

18--On the basis of his belief that tests indicated the practicability of ship-based helicopters, the Chief of Naval Operations separated the pilot training from test and development functions in the helicopter program. He directed that, effective 1 January 1944, a helicopter pilot training program be conducted by the U.S. Coast Guard at Floyd Bennett Field, under the direction of the Deputy Chief of Naval Operations (Air).

20--The Naval Air Training Command was established at Pensacola, Fla., to coordinate and direct, under the Chief of Naval Operations, all naval aviation training in the activities of the Primary, Intermediate, and Operational Training Commands.

20--Two Catalinas of Patrol Squadron 43, at Attu, flew the first Navy photo reconnaissance and bombing mission over the Kuriles.

20--Commander Frank A. Erickson, USCG, reported that Coast Guard Air Station, Floyd Bennett Field had experimented with a helicopter used as an airborne ambulance. An HNS-1 helicopter made flights carrying, in addition to its normal crew of a pilot and a mechanic, a weight of 200 pounds in a stretcher suspended approximately 4 feet beneath the float landing gear. In further demonstrations early the following year, the stretcher was attached to the side of the fuselage and landings were made at the steps of the dispensary.

25--Aircraft from a two-carrier task group (Rear Admiral F. C. Sherman) attacked shipping at Kavieng, New Ireland, as a covering operation for landings by the Marines in the Borgen Bay area of New Britain on the following day.

31--Fleet Air Wing 17 departed Australia and set up headquarters at Samarai on the tip of the Papuan Peninsula of New Guinea.

Appendix I

The Evolution of Aircraft Carriers

1944

JANUARY

3--Helicopter Mercy Mission--Commander Frank A. Erickson, USCG, flying an HNS-1 helicopter, made an emergency delivery of 40 units of blood plasma from lower Manhattan Island to Sandy Hook where the plasma was administered to survivors of an explosion on the destroyer *Turner* (DD 648). In this, the first helicopter lifesaving operation, Commander Erickson took off from Floyd Bennett Field, flew to Battery Park on Manhattan Island to pick up the plasma and then to Sandy Hook. The flight was made through snow squalls and sleet which grounded all other types of aircraft.

11--The first U.S. attack with forward-firing rockets was made against a German U-boat by two TBF-1C's of Composite Squadron 58 from the escort carrier Block Island.

16--Lieutenant (jg) S. R. Graham, USCG, while en route from New York to Liverpool in the British freighter *Daghestan* made a 30 minute flight in an R-4B (HNS-1) from the ship's 60 by 80 foot flight deck. Weather during the mid-winter crossing of the North Atlantic permitted only two additional flights and, as a result, the sponsoring Combined Board for Evaluation of the Ship-based Helicopter in Anti-Submarine Warfare concluded that the helicopter's capability should be developed in coastal waters until models with improved performance became available.

18--Catalinas of VP-63, based at Port Lyautey, began barrier patrols of the Strait of Gibraltar and its approaches with Magnetic Airborne Detection (MAD) gear and effectively closed the Strait to enemy U-boats during daylight hours until the end of the war.

29 January-22 February--Occupation of the Marshall Islands--Six heavy and six light carriers, in four groups of Task Force 58 (Rear Admiral M. A. Mitscher), opened the campaign to capture the Marshalls (29 Jan) with heavy air attacks on Maloelap, Kwajalein, and Wotje. On the first day the defending enemy air forces were eliminated and complete control of the air was maintained by carrier aircraft during the entire operation. Eight escort carriers, attached to the Attack Forces of the Joint Expeditionary Force, arrived in the area early the morning of D-day. Aircraft from the carriers flew cover and antisubmarine patrols for attack shipping and assisted two fast carrier groups, providing air support for landings on Kwajalein and Majuro Atolls (31 Jan), Roi and Namur (1 Feb), and for operations ashore. The AGC command ship, used for the first time during this campaign, provided greatly improved physical facilities for the Support Air

Appendix I

The Evolution of Aircraft Carriers

Commander. Here, the Support Air Commander first assumed control of Target Combat Air Patrol, previously vested in carrier units, and a Force Fighter Director on his staff coordinated fighter direction. Two fast carrier groups to the west kept Eniwetok Atoll neutralized until the initial objectives were achieved. Their early achievement permitted the second phase of the campaign, Seizure of Eniwetok, earlier than the planned date of 10 May. The landings (17 Feb) and the ground action were supported by aircraft from one fast carrier group and one escort carrier group. Covering operations were provided by the First Strike on Truk (17-18 Feb), carried out by the Truk Striking Force (Vice Admiral R. A. Spruance), built around three fast carrier groups. In a 2-day attack, the carriers launched 1,250 combat sorties against this key naval base and exploded the myth of its impregnability with 400 tons of bombs and torpedoes, sinking 37 war and merchant ships aggregating 200,000 tons and doing heavy damage to base installations. In this action the first night bombing attack in the history of U.S. carrier aviation was carried out by VT-10 from the Enterprise with 12 radar equipped TBF-1C's. The attack, delivered at low level, scored several direct hits on ships in the harbor. In a brief enemy air attack on the same night, Intrepid was hit by an aerial torpedo. For the campaign, night fighter detachments of VF(N)-76 and VF(N)-101 (assigned F6F-3's and F4U-2's equipped with AIA radar) were assigned to five carriers and, while not widely used, were on occasion vectored against enemy night raiders.

30--To effect the neutralization of Wake Island during the Marshalls operation, two squadrons of Coronados from Midway Island made the first of four night bombing attacks. Repetitions of the 2,000-mile round trip mission were completed on 4, 8, and 9 February.

FEBRUARY

2--The last of the World War II ceilings for Navy aircraft, calling for an increase to 37,735 useful planes, was approved by the President.

3--Flight Safety Bulletin No. 1 was issued jointly by the Deputy Chief of Naval Operations (Air) and the Chief of the Bureau of Aeronautics, announcing their intention to issue consecutively numbered bulletins concerning the safe operation of naval aircraft.

4--In a test of refueling operations with the CVE *Altamaha* off San Diego, the K-29 of Blimp Squadron 31 made the first carrier landing by a non-rigid airship.

4--The first photo reconnaissance of Truk was made by 2 PB4Y's of VMD-254 on a 12-hour night flight from the Solomon Islands. Cloud cover prevented complete coverage but the information acquired was useful in planning the carrier strike which hit later in the month.

Appendix I

The Evolution of Aircraft Carriers

15--A new command, Forward Area, Central Pacific, was established to control the operations of shore based air forces and naval forces assigned to the Ellice, Gilbert, and Marshall Islands.

20--On completion of the strike on Truk, a small unit composed of the *Enterprise*, one cruiser, and six destroyers (Rear Admiral J. W. Reeves) separated from the main force and launched two air strikes on Jaluit.

23--Two carrier groups of Task Force 58 (Rear Admiral M. A. Mitscher), after successfully fending off a series of determined enemy air attacks during the night, hit targets on Saipan, Tinian, Rota, and Guam for the dual purpose of reducing enemy air strength in the Marianas and to gather photo intelligence for the impending invasion. The combined efforts of pilots and antiaircraft gunners accounted for 67 enemy aircraft shot down and 101 destroyed on the ground.

24--The first detection of a submerged enemy submarine by the use of MAD gear was made by Catalinas of VP-63, on a MAD barrier patrol of the approaches to the Strait of Gibraltar. They attacked the U-761 with retrorockets, and with the assistance of two ships and aircraft from two other squadrons, sank it.

MARCH

4--A reduction in flight training was visualized as the total outputs for 1944, 1945, and 1946 were fixed at 20,500, 15,000 and 10,000 pilots respectively.

6--A new specification for color of naval aircraft went into effect. The basic camouflage scheme, used with fleet aircraft, was modified slightly to provide for use of non-specular sea blue on upper fuselage surfaces; airfoil surfaces visible from above remained semi-gloss sea blue and other surfaces visible from below, semi-gloss insignia white. A new basic non-camouflage color scheme, all aluminum, was specified for general use on aircraft not in the combat theater. The maximum visibility color scheme used on primary trainers became glossy orange yellow overall.

15--The twin-engine North American Mitchell, PBJ, was taken into combat for the first time in its naval career in an attack on Rabaul by pilots of Marine Bombing Squadron 413.

18--Task Group 50.10 (Rear Admiral W. A. Lee), composed of the Lexington, two battleships, and a destroyer screen, bombed and bombarded bypassed Mili in the Marshalls.

Appendix I

The Evolution of Aircraft Carriers

20--Two escort carriers provided cover and airspot for the battleship and destroyer bombardment of Kavieng and nearby airfields in a covering action for the occupation of Emirau.

22--A new specification for color of fighter aircraft went into effect. It directed that fighters be painted glossy sea blue on all exposed surfaces.

26--Corsairs of VMF-113 from Engebi flew the first fighter escort for AAF B-25'S on the 360 mile bombing mission against Ponape, and were so effective in destroying enemy interceptors that later missions over the island were unmolested.

27--*Saratoga* (Captain J. H. Cassady) and three destroyers, assigned to temporary duty with the Royal Navy, joined the British Eastern Fleet in the Indian Ocean approximately 1,000 miles south of Ceylon.

30 March-1 April--Strikes on the Western Carolines--In an operation designed to eliminate opposition to the landings at Hollandia and to gather photo intelligence for future campaigns, a strong Fifth Fleet force, built around Task Force 58 (Vice Admiral M. A. Mitscher) with 11 carriers, launched a series of attacks on Palau, Yap, Ulithi, and Woleai, and shipping in the area. Aerial mining of Paliu Harbor by Torpedo Squadrons 2, 8, and 16, was the first such mission by carrier aircraft and the first large scale daylight mining operation of the Pacific war. The attacks accounted for 157 enemy aircraft destroyed, 28 ships of 108,000 tons sunk, and denial of the harbor to the enemy for an estimated 6 weeks.

APRIL

15--Air-Sea Rescue Squadrons (VH) were formed in the Pacific Fleet to provide rescue and emergency services as necessary in the forward areas. Prior to this time the rescue function was performed as an additional duty by regularly operating patrol squadrons.

16--Carrier Transport Squadron, Pacific, was established for administrative and operational control over escort carriers assigned to deliver aircraft, spare parts, and aviation personnel in direct support of Pacific Fleet Operations.

18--In preparation for the campaign to occupy the Marianas, photo-equipped Liberators of VD-3 obtained complete coverage of Saipan, Tinian, and Aguijan Islands. For the 13-hour flight from Eniwetok and return, B-24's of the AAF flew escort for the photo planes and bombed the islands in a diversionary action. This was the first mission by shore-based aircraft over the Marianas.

Appendix I

The Evolution of Aircraft Carriers

19--Saratoga, operating with the British Eastern Fleet, participated in the carrier strike on enemy installations at Sabang in the Netherlands East Indies.

21-24--Landings at Hollandia--Task Force 58 (Vice Admiral M. A. Mitscher) supported the landings of Southwest Pacific Forces in the Hollandia-Aitape section of the north New Guinea coast. The force of five heavy and seven light carriers organized in three groups, launched preliminary strikes on airfields around Hollandia and at Wakde and Sawar (21 April), covered the landings (22 April) at Aitape, Tanahmerah Bay, and Humboldt Bay, and supported troop movements ashore (23-24 April). Eight escort carriers of Task Force 78 (Rear Admiral R. E. Davison) flew cover and antisubmarine patrols over ships of the Attack Group during the approach and provided support for the amphibious assault at Aitape. Carrier aircraft accounted for the destruction of 30 enemy aircraft in the air and 103 on the ground.

23--VR-3 operated the first regularly scheduled NATS transcontinental hospital flight between Washington, D.C., and March Field, Calif.

26--Headquarters of Fleet Air Wing 4 was established on Attu, western most island of the Aleutians.

29 April-1 May--Second Carrier Strike on Truk--Task Force 58 (Vice Admiral M. A. Mitscher), returning to Majuro from the Hollandia operation, launched a 2-day attack on enemy installations and supply dumps at Truk. In addition to damage ashore, three small ships were sunk and 145 enemy aircraft destroyed. Task Group 58.1 (Rear Admiral J. J. Clark), detached from the main force on the second day, flew protective cover for a cruiser bombardment of Satawan, and on 1 May supported bombardment of Ponape with air cover and bombing and strafing attacks.

MAY

1--The command Aircraft, Central Pacific, was dissolved and its functions assumed by Commander Marshalls Sub-Area.

4--A board headed by Rear Admiral A. W. Radford and known by his name, submitted a report that had a direct effect on aviation planning during the latter part of the war and, with modifications to fit the needs of peacetime, extended its influence long after the war. The, Integrated Aeronautic Program for Maintenance, Material and Supply, which evolved from its recommendations, was essentially a plan involving the assignment of new planes to combat units; return of aircraft to the United States for reconditioning and reassignment after specified combat tours; the retirement of second tour aircraft before maintenance

Appendix I
The Evolution of Aircraft Carriers

became costly; and the support of the aeronautical organization through the use of factors and allowances for pools, pipelines, and reconditioning kept realistic by frequent appraisal.

8--The seaplane tender Kenneth Whiting, first of four ships of the class, was commissioned at Tacoma, Wash., Commander R. R. Lyons in command.

8--Commander Naval Forces, Northwest African Waters, approved the assignment of nine Naval Aviators from Cruiser Scouting Squadron 8 (VCS-8) to the 111th Tactical Reconnaissance Squadron (TRS) of the 12TH Army Air Force for flight training and combat operations in North American P-51C Mustangs. Previous combat experience with Curtiss SOC Seagulls and Vought OS2U Kingfishers being used in air spotting and reconnaissance missions proved both types were vulnerable to enemy fighters and antiaircraft fire. The higher performance of fighters such as the P-51 was expected to result in a reduction of casualties on these missions. A total of 11 Naval Aviators participated in combat operations from the cockpits of P-51s while assigned to the 111TH TRS in support of the campaign in Italy and the invasion of southern France. On 2 September 1944 all Naval Aviators assigned to the 111TH returned to their ships, ending a four month long association between the 111TH TRS and VCS-8.

13--To distinguish between fixed and rotary wing heavier-than-aircraft, the helicopter class designation VH plus a mission letter (i.e. VHO for observation and VHN for training) was abolished and helicopters were established as a separate type designated H. The previous mission letters thus became classes designated O, N, and R for observation, training and transport respectively.

13--To meet the needs of the fleet for aviation personnel trained in the use of electronics countermeasures equipment, the Chief of Naval Operations directed that on 1 June, or as soon thereafter as practicable, the Chief of Naval Air Technical Training establish a school to be known as Special Projects School for Air, located initially at NAAS, San Clemente Island, Calif.

15--The first of 16 special transatlantic flights was made by NATS aircraft to the United Kingdom to deliver 165,000 pounds of minesweeping gear essential to the safety of assault shipping during the Normandy invasion. The delivery was successfully completed 23 May.

17--The Bureau of Aeronautics authorized CGAS Floyd Bennett Field to collaborate with the Sperry Gyroscope Company in making an automatic pilot installation in a HNS-1 helicopter.

Appendix I

The Evolution of Aircraft Carriers

17--*Saratoga* participated in the carrier air strike of the British Eastern Fleet on the Japanese base at Soerabaja, Java.

19-20--Third Raid on Marcus--Planes from a three-carrier task force (Rear Admiral A. E. Montgomery) hit Marcus with a predawn fighter sweep and strafed and bombed the island for 2 consecutive days.

23--Third Wake Raid--Carrier Task Group 58.6 (Rear Admiral A. E. Montgomery) shifted from Marcus to hit Wake with five composite bombing, strafing and rocket strikes.

29--The only U.S. carrier lost in the Atlantic, Block Island, was torpedoed and sunk by a German U-boat while engaged in hunter-killer operations in the Azores area.

31--Commander Training Task Force was directed to establish on 1 June, within his command at NAS Traverse City, Michigan, a detachment to be known as Special Weapons Test and Tactical Evaluation Unit to conduct such tests of special weapons and other airborne equipment as were assigned.

JUNE

1--Airships of ZP-14, assigned to antisubmarine operations around Gibraltar, completed the first crossing of the Atlantic by non-rigid airships. The flight began 29 May from South Weymouth, Mass., and ended at Port Lyautey, French Morocco, covering a distance of 3,145 nautical miles in 58 hours. Including time for stopovers at Argentia and the Azores, the airships moved their area of operations across the Atlantic in 80 hours.

1--Air Transport Squadron 9 (VR-9) was formed at Patuxent River and VR-12 at Honolulu to function as headquarters and maintenance squadrons for their respective commands, NATS Atlantic and NATS Pacific.

4--Off Cape Blanco, Africa, a hunter-killer group (Captain D. V. Gallery), composed of the escort carrier Guadalcanal, with VC-8 aboard, and five destroyer escorts, carried out a determined attack on the German submarine U-505, forcing it to surface. Boats from the destroyer escort Pillsbury (DD 133) and the carrier reached the submarine before scuttling charges could accomplish their purpose and the U.S. Navy found itself with a prize of war.

5--The Deputy Chief of Naval Operations (Air) reported that Aviation Safety Boards, established in one large command, had in one-quarter of operation reduced the fatal accident rate by 47 percent. He directed the establishment of

Appendix I

The Evolution of Aircraft Carriers

similar boards in other commands outside of advanced combat areas and the appointment of a flight safety officer in each squadron.

6--Allied Invasion of Normandy--Seventeen naval aviators taken from aviation units on battleships and cruisers were assigned to bombardment duty as part of VCS-7. They operated with units of the British Fleet Air Arm and Royal Air Force, flying gunfire spotting missions in RAF Spitfires over the Normandy beaches from D-day until the June 26.

11 June-10 August--Occupation of the Marianas--Task Force 58 (Vice Admiral M. A. Mitscher), built around seven heavy and eight light carriers, opened the campaign to occupy the Marianas Islands with a late afternoon fighter sweep (11 June) that destroyed one-third of the defending air force. In bombing and strafing attacks on shore installations and on shipping in the immediate area on succeeding days, this force prepared the way for the amphibious assault of Saipan (15 June), supported operations ashore with daily offensive missions, kept the area isolated with attacks on airfields and shipping in the Bonin and Volcano Islands to the north (15-16, 24 June, 3-4 July, 4-5 Aug), and successfully defended the operation against an attack by major fleet forces in the Battle of the Philippine Sea (19-20 June). On the first day (19 June) TF 58 repelled a day-long air attack from carriers and shore bases, destroying 402 enemy planes, and the next day (20 June) launched an air attack late in the afternoon on the retreating Japanese Fleet, sinking the carrier *Hiyo* and two fleet oilers.

Air cover for assault and close air support for operations ashore was provided by aircraft from an initial force of 11 escort carriers attached to Attack Forces. A Navy seaplane squadron VP-16, moved into the area (16 June) and began operations from the open sea. Garrison aircraft were ferried in by escort carriers to operate from captured airfields. First to arrive were Marine observation planes of VMO-4 (17 June), AAF P-47's (22 June), and Marine Corps Night Fighter Squadron 532 (12 July). After organized resistance ended on Saipan (9 July), troops landed on Guam (21 July) and on Tinian (24 July). As the campaign neared successful completion, three groups of Task Force 58 left the area temporarily for Strikes on the Western Carolines (25-28 July). Palau, Yap, Ulithi and other islands were taken under attack while photographic planes obtained intelligence of enemy defenses. This done, the groups steamed north for the fourth side of the campaign on the Bonins and Volcanoes. When Guam was secure (10 Aug), carrier aircraft had accounted for 110,000 tons of enemy shipping sunk and 1,223 aircraft destroyed. In this campaign, groups of the fast carrier force retired in turn to advanced fleet bases for brief periods of rest and replenishment, thus initiating a practice that became standard operating procedure during all future extended periods of action.

Appendix I

The Evolution of Aircraft Carriers

12--In the first deployment of a guided missile unit into a combat theater, elements of Special Task Air Group 1 arrived in the Russell Islands in the South Pacific.

24--The Chief of Naval Operations promulgated plans which provided for a drastic reduction in the pilot training program. This required the transfer of some students already in Pre-Flight, and prior stages of training and the retention of enough to maintain a course in Pre-Flight schools expanded to 25 weeks. The program of "deselection" and voluntary withdrawal of surplus students was instituted by the Chief of Naval Air Training early in the next month. The resulting reductions were directly responsible for the discontinuance of the War Training Service Program in August, closing the Flight Preparatory Schools in September and the release of training stations which began in September.

26--The seaplane tender *Currituck*, first of four ships of her class, was commissioned at Philadelphia, Captain W. A. Evans commanding.

29--The Parachute Experimental Division was established at Lakehurst, NJ for research, development, and testing of parachutes and survival gear.

29--Carrier Air Groups were standardized for all commands under the following designations: CVBG, large carrier air group; CVG, medium carrier air group; CVLG, light carrier air group; CVEG, escort carrier air group (Sangamon class); and VC, escort carrier air group (*Long Island, Charger, Bogue, and Casablanca* class).

30--The Naval Aircraft Modification Unit of the Naval Air Material Center, Philadelphia, was relocated at Johnsville, Pa., where facilities for intensified efforts in guided missiles development and quantity modification of service airplanes were available.

JULY

6--A special Air Unit was formed under ComAirLant, with Commander James A. Smith, Officer in Charge, for transfer without delay to Commander Fleet Air Wing 7 in Europe. This unit was to attack German V-1 and V-2 launching sites with PB4Y-1's converted to assault drones.

6--The Bureau of Aeronautics authorized Douglas to proceed with the design and manufacture of 15 XBT2D airplanes. The single-seat dive-bomber and torpedoplane thus initiated, was designed jointly by BuAer and Douglas engineers. Through subsequent development and model redesignation, these aircraft became the prototypes for the AD Skyraider series of attack planes.

Appendix I

The Evolution of Aircraft Carriers

14--To achieve economy of effort and unity of purpose by coordinating all safety functions through a central organization, a Flight Safety Council was established by the joint action of the Deputy Chief of Naval Operations (Air) and the Chief of the Bureau of Aeronautics, to plan, coordinate, and execute flight safety programs.

14--PB4Y Liberators, of VB-109 based at Saipan, made the first strike on Iwo Jima by shore-based planes.

27--Fleet Air Wing 17 headquarters moved to Manus in the Admiralty Islands.

29--In the first successful test of the Pelican guided missile, conducted 44 miles offshore from NAS New York, two of the four launched against the target ship James Longstreet were hits.

29--A detachment of Bombing Squadron 114 (Liberators) from Port Lyautey, was established under British command at Lagens Airfield in the Azores Islands for antisubmarine operations.

31--The Accelerated Field Service Test Unit at Patuxent River was redesignated Service Test and established as a separate department.

AUGUST

5--The Fast Carrier Task Force was reorganized into First and Second Fast Carrier Task Forces, Pacific, commanded by Vice Admiral M. A. Mitscher and Vice Admiral J. S. McCain respectively.

7--Carrier Division 11 was established at Pearl Harbor, Rear Admiral M. B. Gardner commanding. This division, composed of the carriers Saratoga and Ranger, was the first in the U.S. Navy specifically established for night operations.

10--Naval Air Bases commands were established within each Naval District, the Training Command, and for Marine Corps Bases, and were charged with the military direction and administrative coordination of matters affecting the development and operational readiness of aviation facilities in their respective areas.

10--The operating aircraft complement of Carrier Air Groups was revised to 54 VF, 24 VB and 18 VT with the provision that four night fighters and two photo planes be included among the 54 VF.

Appendix I

The Evolution of Aircraft Carriers

11--An electric powered rescue hoist was installed on an HNS-1 helicopter at CGAS Floyd Bennett Field. During the ensuing 4 day test period, in which flights were conducted over Jamaica Bay, the feasibility of rescuing personnel from the water and of transferring personnel and equipment to and from underway boats was demonstrated. Six weeks later, a hydraulic hoist, which overcame basic disadvantages of the electric hoist, was installed and successfully tested, leading to its adoption for service use.

11--Dr. M. F. Bates of the Sperry Gyroscope Company submitted a brief report of the trial installation and flight test of a helicopter automatic pilot (cyclic pitch control) in an HNS-1 at CGAS Floyd Bennett Field.

15-29--Landings in Southern France--Two United States and seven British escort carriers of the Naval Attack Force (Rear Admiral T. H. Troubridge, RN) supplied defensive fighter cover over the shipping area, spotted for naval gunfire, flew close support missions, made destructive attacks on enemy concentrations and lines of communication and otherwise assisted Allied troops landing between Toulon and Cannes and advancing up the Rhone Valley.

20-23--The nonrigid airship K-111, under command of Lieutenant Commander F. N. Klein, operating in conjunction with the Escort Carrier Makassar Strait off San Diego, demonstrated the feasibility of refueling and replenishing airships from aircraft carriers. In this operation of 72.5 hours duration, the airship's crew was relieved every 12 hours and its engines were operated continuously. In one evolution, the airship remained on deck for 32 minutes.

24--The first night carrier air group, CVLG(N)-43, was established at Charlestown, R.I. Its component squadrons VF (N)-43 and VT(N)-43, the latter the first of the night torpedo squadrons, were established the same day.

24--Fleet Air Wing 10 moved forward from Perth, Australia, to Los Negros in the Admiralty Islands, to support the advance of Southwest Pacific Forces on the Philippines.

31 August-30 September--Occupation of Palau and Morotai--Simultaneous landings by Central and Southwest Pacific Forces were preceded by wide-flung operations of four carrier groups of Task Force 38 (Vice Admiral M. A. Mitscher), which committed only part of its strength in direct support and operated principally in covering action. TG 38.4 (Rear Admiral R. E. Davison) opened the campaign with attacks on the Bonin and Volcano Islands (31 Aug-2 Sep). The entire Fast Carrier Force hit the Palau area (6-8 Sep), leaving TG 38.4 to maintain the neutralization of Palau, and moved against the Philippines with fighter sweeps over Mindanao airfields (9-10 Sep) and strikes in the Visayas

Appendix I

The Evolution of Aircraft Carriers

(12-14 Sep). Here TG 38.1 (Vice Admiral J. S. McCain) separated to hit Mindanao (14 Sep) and to support Landing on Morotai by Southwest Pacific Forces (15 Sep). The landings were preceded by bombing and strafing attacks and were supported (15-16 Sep) by TG 38.1 aircraft and additionally by six escort carriers of TG 77.1 (Rear Admiral T. L. Sprague). Landings on Peleliu by Central Pacific Forces (15 Sep) were preceded by preliminary carrier air attacks (12-14 Sep) from TG 38.4 and from four CVE's of Carrier Unit One (Rear Admiral W. D. Sample). Continued support was given by the same fast carrier group (15-18 Sep) and until the end of the month by a total of 10 escort carriers operating in TG 32.7 (Rear Admiral R. A. Ofstie). Carrier air support was also provided for Landings on Agaur (17 Sep), Ulithi (23 Sep), and the shore-to-shore movement from Peleliu to Ngesebus (28 Sep) support for the latter including strikes by Marine Corps land-based units from Peleliu, the first of which, VMF(N)-541, had arrived 24 September. Following the action at Morotai, TG 38.1 re-joined the main body of Fast Carriers which then launched strikes on airfields and shipping around Manila (21-22 Sep) and hit airfields, military installations, and shipping in the central Philippines (24 Sep) before retiring. In this month of action, carrier planes destroyed 893 enemy aircraft and sank 67 war and merchant ships totalling 224,000 tons. Enemy weakness in the central Philippines, uncovered by carrier air action, changed plans for reentry into the Philippines, shifting the assault point from southern Mindanao to Leyte and advancing the assault date from mid-November to 20 October.

SEPTEMBER

1--Project Bumblebee (as it was later known) came into being as the Bureau of Ordnance reported that a group of scientists from Section T of the Office of Scientific Research and Development were investigating the practicability of developing a jet-propelled, guided, anti-aircraft weapon. Upon completion of the preliminary investigation, a developmental program was approved in December by the Chief of Naval Operations. In order to concentrate upon the guided missile phase of the anti-aircraft problem, the OSRD and Applied Physics Laboratory of Johns Hopkins University, completed withdrawal, also in December, from the proximity fuze program which thus came completely under the Bureau of Ordnance.

3--Lieutenant Ralph Spaulding of Special Air Unit, Fleet Air Wing 7, flew a torpex-laden drone Liberator from an airfield at Feresfield, England, set radio control and parachuted to ground. Ensign J. M. Simpson, controlling the Liberator's flight from a PV, sought to hit submarine pens on Helgoland Island; however, he lost view of the plane in a rain shower during the final alignment and relying only upon the drone's television picture of the terrain hit the barracks and industrial area of an airfield on nearby Dune Island.

Appendix I

The Evolution of Aircraft Carriers

3--Fourth Wake Raid--A strike group of one carrier, with cruisers and destroyers, hit enemy positions on Wake.

6--A contract was awarded to McDonnell Aircraft Corporation for development of the Gargoyle or LBD-1, a radio controlled low-wing gliding bomb fitted with a rocket booster and designed for launching from carrier-based dive-bombers and torpedo planes against enemy ships.

6--As the scope of the aviation safety program was enlarged, a Flight Safety Section was established in the Office of the Deputy Chief of Naval Operations (Air), and was assigned the direction and supervision of the aviation safety program.

9--Fleet Air Wing 17 moved forward to the Schouten Islands to direct patrol plane operations supporting the occupation of Morotai by Southwest Pacific Forces.

11--Commander Fleet Air Wing 1, based on Hamlin, transferred from Espiritu Santo in the South Pacific to Guam to direct the operations of patrol squadrons in the Central Pacific.

18--The Pelican guided missile production program was terminated and the project returned to a developmental status. Despite reasonably successful during the preceding 6 weeks, this decision was made because of tactical, logistic and technical problems involved in its use.

27--Guided Missiles were used in the Pacific as Special Task Air Group 1, from its base on Stirling in the Treasury Islands, began a combat demonstration of the TDR assault drone. The drones had been delivered to the Russell Islands by surface shipping and flown 45 miles to bases in the Northern Solomons where they were stripped for pilotless flight and armed with bombs of up to 2,000 pounds. For combat against heavily defended targets, a control operator in an accompanying TBM guided the drone by radio and directed the final assault by means of a picture received from a television camera mounted in the drone. In the initial attack, against antiaircraft emplacements in a beached merchant ship defending Kahili airstrip on South Bougainville, two out of four TDR's struck the target ship.

Appendix I

The Evolution of Aircraft Carriers

OCTOBER

1--Patrol Squadrons (VP) and multi-engine bombing squadrons (VB) were renamed and re-designated patrol bombing squadrons (VPB).

7--A new Bureau of Aeronautics color specification went into effect which provided seven different color schemes for aircraft depending upon design and use. The most basic change was the use of glossy sea blue all over on carrier based aircraft and on seaplane transports, trainers and utility aircraft. The basic non-specular camouflage color scheme, semigloss blue above and nonspecular white below, was to be applied to patrol and patrol bombing types and to helicopters. For antisubmarine warfare, two special camouflage schemes--gray on top and sides and white on bottom, or white all over--were prescribed with the selection dependent upon prevailing weather conditions (this had been used by COMNAVAIRLANT since 19 July 1943). All aluminum was to be used on landplane transports and trainers and landplane and amphibian utility aircraft. Orange-yellow was to be used upon target-towing aircraft and primary trainers. Another new scheme, glossy red, was specified for target drones.

7--Provision was made for the optional use by tactical commanders of special identification markings on combat aircraft, such markings preferably to be applied with temporary paint.

10 October-30 November--Occupation of Leyte-- The opening blow of the campaign was struck (10 Oct) by Task Force 38 (Vice Admiral M. A. Mitscher) against airfields on Okinawa and the Ryukyus. This force, built around 17 carriers hit airfields on northern Luzon (11 and 14 Oct), on Formosa (12-14 Oct), and in the Manila area (15 Oct), destroying 438 enemy aircraft in the air and 366 on the ground in 5 strike days. These and other strikes concentrated on reinforcement staging areas and effectively cleared the air for the landing (20 Oct) of Southwest Pacific Army troops on Leyte. Fast carrier support of the ground campaign was supplemented (18-23 Oct) by the action of 18 CVE's organized in three elements under TG 77.4 (Rear Admiral T. L. Sprague).

A major disruptive effort by the Japanese Fleet was opposed by surface and air elements of the Seventh Fleet (Vice Admiral T. C. Kinkaid) and by the Fast Carrier Force of the Third Fleet (Vice Admiral W. F. Halsey) in three related actions of The Battle for Leyte Gulf (23-26 Oct). As the Japanese Fleet, in three elements identified as Southern, Central, and Northern Forces, converged on Leyte Gulf from as many directions, Fast Carrier Force aircraft (24 Oct) hit the Southern Force in the Sulu Sea, attacked the Central Force in the Sibuyan Sea, sinking the 63,000 ton battleship *Musashi* and a destroyer, and was itself under air attack resulting in the loss of *Princeton*. Seventh Fleet surface elements turned back the Southern Force in a brief intensive action before daylight in the

Appendix I

The Evolution of Aircraft Carriers

Battle of Surigao Strait (25 Oct), sinking two battleships and three destroyers. The Japanese Central Force made a night passage through San Bernardino Strait and at daylight took under fire six escort carriers and screen of TG 77.4, and was opposed by a combined air and ship action in the Battle Off Samar (25 Oct) in which *Gambier Bay*, two destroyers, and one destroyer escort were sunk by enemy gunfire and three Japanese heavy cruisers were sunk by carrier air. At the same time the Fast Carrier Force met the Northern Force in the Battle Off Cape Engano, sinking the heavy carrier *Zuikaku* and light carriers *Chiyoda, Zuiho, and Chitose*, the latter with the assistance of cruiser gunfire. Off Leyte, Kamikaze pilots, in the first planned suicide attacks of the war, hit the escort carriers and sank the St. Lo and damaged the *Sangamon, Suwannee* (AO 33), *Santee, White Plains, Kalinin Bay, and Kitkun Bay*. As remnants of the Japanese Fleet limped homeward through the Central Philippines, (26-27 Oct) carrier aircraft sank a light cruiser and four destroyers to bring Japanese battle losses to 26 major combatant ships totaling over 300,000 tons.

Direct air support in the Leyte-Samar area was assumed by Allied Air based at Tacloban (27 Oct) and 2 days later the escort carriers retired. Later one group operated at sea to protect convoys from the Admiralties against air and submarine attack (19-28 Nov) and another group performed the same services (14-23 Nov) for convoys from Ulithi. The Fast Carrier Force also continued support for 2 days attacking airfields on Luzon and in the Visayas (27-28 Oct), shipping near Cebu (28 Oct), and Luzon airfields and shipping in Manila Bay (29 Oct). In supporting operations during October, carrier aircraft destroyed 1,046 enemy aircraft.

Requirements for continued carrier air support for the campaign caused cancellation of a planned Fast Carrier Strike on Tokyo, and Task Force 38 (now under Vice Admiral J. S. McCain) sortied from Ulithi to hit Luzon and Mindoro airfields and strike shipping in Manila Bay (5-6 Nov), sinking a heavy cruiser and other ships; hit a reinforcement convoy of four transports and five destroyers in Ormoc Bay (11 Nov) sinking all but one destroyer; shifted to Luzon and the Manila area (13-14 Nov) and sank a light cruiser, four destroyers, and 20 merchant and auxiliary ships; hit the same areas again (19 and 25 Nov), sinking another heavy cruiser and several auxiliaries; and wound up the month's operations with an aerial score of 770 enemy aircraft destroyed. During these actions, the force was under several Kamikaze attacks which damaged the carriers Intrepid (29 Oct), Franklin and Belleau Wood (30 Oct), Lexington (5 Nov), Essex, Intrepid, and Cabot (25 Nov), two seriously enough to require Navy Yard repairs.

14--The Amphibious Forces Training Command, Pacific, was directed to form mobile Air Support Training Units to train Carrier Air Groups and Marine Corps squadrons in the technique of close air support operations.

17--Commander Fleet Air Wing 10, on Currituck, arrived in Philippine waters and directed patrol plane operations in support of the occupation of Leyte.

19--Commander Fleet Air Wing 17 moved to Morotai, N.E.I., to support Southwest Pacific operations against the Philippines.

21--A new command, Marine Carrier Air Groups, was established under Aircraft FMFPac to direct the formation and training of Marine Corps squadrons destined to operate from air support escort carriers. Current plans called for the formation of six Marine Carrier Air Groups, each composed of a fighter and a torpedo squadron, four of them to be assigned to escort carriers and two to function as replacement and training groups.

25--In recognition of the difference in functions performed, Carrier Aircraft Service Units and Patrol Aircraft Service Units, operating at advanced bases, were redesignated Combat Aircraft Service Units (Forward), short title CASU(F), while those in the continental United States and Hawaii retained the original title.

26--The last attack in a month long demonstration of the TDR assault drone was made by Special Task Air Group, thereby concluding the first use of the guided missile in the Pacific. During the demonstration a total of 46 drones were expended of which 29 reached the target areas: two attacked a lighthouse on Cape St. George, New Ireland, making one hit which demolished the structure; nine attacked anti-aircraft emplacements on beached ships achieving six direct hits and two near misses; and 18 attacked other targets in the Shortlands and Rabaul areas making 11 hits.

NOVEMBER

6--Recognition of the future importance of turbojet and turboprop powerplants led the Bureau of Aeronautics to request the Naval Air Material Center to study requirements for a laboratory to develop and test gas-turbine powerplants. This initiated action which led to the establishment of the Naval Air Turbine Test Station, Trenton, N.J.

17--The Bureau of Aeronautics reported that technical studies were underway to determine the feasibility of launching an adaptation of the JB-2, a U.S. Army version of the German V-l Buzz Bomb, from escort carriers for attacks on enemy surface vessels and shore targets. Modifications visualized included installation of radio controls and a radar beacon. As subsequently developed, this became the Loon.

23--Training Task Force Command was dissolved and its facilities, personnel, and equipment reallocated.

27--*Commencement Bay*, first of her class built from the last U.S. escort carrier design, was commissioned at Tacoma, Washington, Captain R. L. Bowman commanding.

29--The changing character of the war was reflected in a revision of the aircraft complement of Essex Class Carrier Air Groups to 73 VF, 15 VB and 15 VT. The fighter complement was to be filled by two squadrons of 36 planes each

Appendix I

The Evolution of Aircraft Carriers

plus one for the Air Group Commander and to include four VF(N), two VF(P) and two VF(E). The change to the new figures was gradual, beginning with the assignment of Marine fighter squadrons in December and continued with the establishment of VBF squadrons the following month.

30--Fleet Air Wing 10 headquarters became shore based on Jinamoc Island in the Philippines.

DECEMBER

1--Electronics Tactical Training Unit was established at NAS Willow Grove to train personnel of the Airborne Coordinating Group as instructors in the operation of all newer types of airborne electronics apparatus including search, navigation, identification, and ordnance radar.

7--Chourre was commissioned as the first aviation repair ship of the U.S. Navy, Captain A. H. Bergeson commanding.

11--The steady decline in U-boat activity in the Caribbean during the year permitted a reduction of blimp operations over the southern approaches, and Fleet Airship Wing 5 at Trinidad was disestablished.

12--Three Evacuation Squadrons (VE) were established in the Pacific from Air Sea Rescue Squadron elements already providing evacuation services.

13--Escort Carrier Force, Pacific (Rear Admiral C. T. Durgin), was established for administrative control over all escort carriers operating in the Pacific, excepting those assigned to training and transport duty.

14-16--Support of the Landings on Mindoro-- Six escort carriers of Task Unit 77.12.1 (Rear Admiral F. B. Stump) and Marine Corps shore-based air flew cover for the passage of transport and assault shipping through the Visayas (12-14 Dec). The escort carriers provided direct support for landings by Army troops (15 Dec) and in the assault area (16-17 Dec). On the night of D-day Navy seaplanes joined with operations from Mangarin Bay. The covering support of Task Force 38 (Vice Admiral J. S. McCain), with seven heavy and six light carriers, began with fighter sweeps over Luzon airfields (14 Dec) and continued with successive combat air patrols relieved on station, which spread an aerial blanket over Luzon (14-16 Dec) and effectively pinned down all enemy aircraft on the island, and accounted for a major share of the 341 enemy aircraft destroyed in the brief campaign.

18--Third Fleet units, refueling east of the Philippines, were overtaken by an unusually severe typhoon which formed nearby. Three destroyers capsized in the high seas and several ships were damaged, including four light carriers of Task Force 38 and four escort carriers of the replenishment group. 28--Marine Corps Fighter Squadrons 124 and 213, the first to operate from fast carriers in combat, reported for their first tour of carrier duty aboard Essex in Uluthi.

30--The specification on aircraft color was amended to provide that patrol and patrol bombing landplanes received a color scheme that was in general similar to that prescribed for carrier based airplanes. Specifically, the patrol and patrol bombers were to be painted semi-gloss sea blue on top and bottom surfaces of wings and on all horizontal tail surfaces; other tail surfaces and the fuselage were to be non-specular sea blue.

1945

JANUARY

1--Carrier Training Squadron, Pacific, composed of two carrier divisions, was established in the Pacific Fleet to provide operational control over carriers employed in training Carrier Air Groups out of Pearl Harbor and San Diego.

2--Eighteen Fighter Bomber Squadrons (VBF) were established within existing Carrier Air Groups to adjust their composition to the needs of changed combat requirements in the Pacific.

2--Headquarters of Fleet Air Wing 17, based on Tangier, directed patrol plane support of the Lingayen Gulf operations from San Pedro Bay.

3-22--Invasion of Luzon--Southwest Pacific Force operations against Luzon were directly supported by Seventh Fleet escort carriers in Task Group 77.4 (Rear Admiral C. T. Durgin) and indirectly by the fast carriers in Task Force 38 (Vice Admiral J. S. McCain) of Third Fleet and Central Pacific Forces. Task Group 77.4, with 17 escort carriers, covered the approach of the Luzon Attack Force against serious enemy air opposition from Kamikaze pilots which sank Ommaney Bay (4 Jan), and damaged several ships including escort carriers Manila Bay and Savo Island (5 Jan). It conducted preliminary strikes in the assault area (7-9 Jan), covered the landings in Lingayen Gulf (9 Jan), and supported the inland advance of troops ashore (9-17 Jan). Among the ships damaged by Kamikaze pilots opposing the landings were the escort carriers Kadashan Bay and Kitkun Bay (8 Jan), and Salamaua (13 Jan). Task Force 38, with seven heavy and four light carriers in three groups and one heavy and one light carrier in a night group, and accompanied by a Replenishment Group with one hunter-killer and seven escort carriers, concentrated on the destruction of enemy air power and air installations in surrounding areas. In spite of almost continuous bad weather which hampered flight operations during the entire month, this force launched offensive strikes on Formosa and the Ryukyus (3-4 Jan), a 2-day attack on Luzon (6-7 Jan) and on fields in the Formosa-Pescadores-Ryukyus area (9 Jan), destroying over 100 enemy aircraft and sinking 40,000 tons of merchant and small combatant ships in 1 week of preliminary action. During the night (9-10 Jan) Task Force 38 made a high-

Appendix I
The Evolution of Aircraft Carriers

speed run through Luzon Strait followed by the Replenishment Group which passed through Balintang Channel, for Operations in the South China Sea (9-20 Jan). Strikes (12 Jan), over 420 miles of the Indo-China coast, reached south to Saigon and caught ships in the harbor and in coastal convoys with devastating results, sinking 12 tankers, 20 passenger and cargo vessels and numerous small combatant ships, totaling 149,000 tons. Moving northward to evade a typhoon, the force hit targets at Hong Kong, the China Coast, and Formosa (15 Jan) and next day concentrated on the Hong Kong area damaging enemy shore installations and sinking another 62,000 tons of shipping. As inclement weather persisted, the force left the South China Sea with an after dark run through Balintang Channel (20 Jan) and hit Formosa, the Pescadores, and Okinawa against enemy air opposition which damaged the Ticonderoga and Langley (20 Jan) and repeated the attack in the Ryukyus next day to finish off 3 weeks of action with an aerial score of over 600 enemy aircraft destroyed and 325,000 tons of enemy shipping sunk.

11--The Bureau of Ordnance assigned the first task on Project Bumblebee to the Applied Physics Laboratory, thus formally establishing the program for development of a ram-jet powered, guided, anti-aircraft weapon from which the Talos, Terrier, and Tartar missiles eventually emerged.

29-31--Six escort carriers of Task Group 77.4 (Rear Admiral W. D. Sample) provided air cover and support for landings by Army troops at San Antonio near Subic Bay (29 Jan), on Grande Island in the same area (30 Jan) and at Nasugbu, south of the entrance to Manila Bay (31 Jan).

FEBRUARY

6--The Chief of Naval Operations directed that, following a period of training at NAS Kaneohe Bay, VPB Squadrons 109, 123, and 124 of Fleet Air Wing 2 be equipped to employ the SWOD Mark 9 Bat glide bomb in combat.

15--The West Coast Wing of the Naval Air Transport Service was disestablished and its squadrons reassigned to the Pacific and Atlantic Wings.

16 February-16 March--Capture of Iwo Jima--The Marine Corps assault of 19 February was preceded and supported by two separate carrier elements of the Central Pacific Force. The first of these was Task Force 58 under Vice Admiral Mark A. Mitscher, the second was Task Group 52.2 under Rear Admiral C.T. Durgin.
On 16-17 February Mitscher moved against Japan with nine heavy and five light carriers in four groups, and two heavy carriers in a night group. Carrier aircraft hit Japanese air bases in the Tokyo plains. From 19 to 23 February, his forces

Appendix I

The Evolution of Aircraft Carriers

supported Marine Corps landings and operations on Iwo Jima and flew neutralization strikes against the Bonins. On 25 February, he returned for a second strike on Tokyo. On 1 March he struck at Okinawa and the Ryukyus and then retired to Ulithi leaving in his wake 648 enemy aircraft destroyed and 30,000 tons of merchant shipping sunk.

Task Group 52.2 began the campaign with nine escort carriers; it was later augmented by two more escorts and one night CV. On 16-18 February, Admiral Durgin carried out air strikes on Iwo Jima's shore defenses to reduce their resistance to the impending Marine Corps landing. From 19 February to 11 March he flew missions in direct support of Marine Corps ground operations and neutralized airstrips in the Bonins.

In counter attacks, the Japanese were not entirely unsuccessful. On 21 February a Kamikaze raid upon Task Group 52 sank the escort carrier Bismarck Sea, seriously damaged Saratoga, and did minor damage to Lunga Point. But new air defense elements in the U.S. fleet were functional and noteworthy; they included the altitude-determining radar on LSTs and a Night Fighter Director on the Air Support Commander's organization.

Other U.S. operations deserve mention. Task Group 50.5, under Commodore D. Ketcham, was based in the Marianas. The Group's shore-based aircraft conducted shipping reconnaissance and air-sea rescue between Japan and Iwo Jima. They also flew offensive screens for carrier raids and expeditionary forces. Similar operations were carried out by patrol planes of Fleet Air Wing 1 from tenders anchored in the lee of Iwo Jima (28 Feb-8 Mar). Marine Corps Observation Squadrons 4 and 5, which arrived on CVEs and on LSTs equipped with Brodie gear, began operations from Iwo Jima airfields on 27 February. Army fighters were flown in from Saipan on 6 March, and Marine Corps Torpedo Squadron 242 arrived on 8 March; they flew day and night combat air patrols and provided all air support upon the departure of the last CVEs on 11 March. Iwo Jima was secured on 16 March.

19--Commander Fleet Air Wing 1 went to sea aboard *Hamlin* to direct patrol squadrons in support of the Iwo Jima campaign and remained in the area until the island was secure.

26--Headquarters of Fleet Air Wing 17 was established ashore at Clark Field on Luzon.

Appendix I

The Evolution of Aircraft Carriers

MARCH

3--The Naval Air Transport Service was reorganized and established as a Fleet Command with headquarters at NAAS Oakland, to operate under the immediate direction of CominCh and CNO.

3--The Naval Air Technical Training Command was incorporated into the Naval Air Training Command.

7--The Commanding Officer, CGAS Floyd Bennett Field reported that a dunking sonar suspended from an XHOS-1 helicopter had been tested successfully.

7--The tandem rotor XHRP-X transport helicopter, built under Navy contract by P-V Engineering Forum made its first flight at the contractor's plant at Sharon Hill, Pennsylvania with Frank N. Piasecki as pilot and George N. Towson as copilot.

8--A rocket powered Gorgon air-to-air missile was launched from a PBY-5A and achieved an estimated speed of 550 m.p.h. in its first powered test flight, conducted off Cape May, N.J. under the direction of Lieutenant Commander M. B. Taylor.

17--Responsibility for evacuating wounded personnel was assigned to the Naval Air Transport Service.

18 March-21 June--The Okinawa Campaign--The last, and for naval forces the most violent of the major amphibious campaigns of World War II, was supported by three separately operating carrier forces, by tender-based patrol squadrons, by Marine and Army air units based in the immediate area and by Army and Navy air units based in other areas. On 28 May a change in overall command from the Fifth Fleet (Admiral R. A. Spruance) to the Third Fleet (Admiral W. F. Hasley) took place, which changed all task number designations from the 50's to the 30's. (In this account, first designations are used throughout.)

The fast carriers of Task Force 58 (Vice Admiral M. A. Mitscher) began the attack. With an original strength of 10 heavy and six light carriers, this force launched neutralization strikes on Kyushu, Japan (18-22 Mar), destroying 482 enemy aircraft by air attack and another 46 by ship's gunfire and began pre-assault strikes on Okinawa (23 Mar). During these preliminaries, Kamikaze pilots, employing conventional aircraft, bombs, and Baka flying bombs (first

Appendix I

The Evolution of Aircraft Carriers

observed on 21 Mar) retaliated with attacks which seriously damaged the carrier Franklin and scored hits on four others.

For the next 3 months the fast carrier force operated continuously in a 60-mile-square area northeast of Okinawa and within 350 miles of Japan, from which position it neutralized Amami Gunto airfields, furnished close air support for ground operations, intercepted enemy air raids, and on occasion moved northward to hit airfields on Kyushu.

Task Group 52.1 (Rear Admiral C. T. Durgin), originally 18 escort carriers strong, conducted pre-assault strikes and supported the occupation of Kerama Retto (25-26 Mar), joined in the pre-assault strikes on Okinawa (27-29 Mar) and, from a fairly restricted operating area southeast of the island, supported the landings and flew daily close support for operations ashore until the island was secure (21 June). The arrival in May of two CVE's with Marine Carrier Air Groups on board marked the combat debut in Marine Air Support carriers.

Task Force 57 (Vice Admiral H. B. Rawlings, RN), a British task force built around four carriers, operated south of Okinawa (26 Mar-20 Apr and 3-25 of May), from which position it neutralized airfields on Sakishima Gunto and Formosa, and intercepted air raids headed for the assault area. Subject to frequent suicide attacks, all four carriers took hits in the course of their action, but all remained operational.

Patrol squadrons of Fleet Air Wing 1, based on seaplane tenders at Kerama Retto, conducted long-range anti-shipping search over the East China Sea to protect assault forces from enemy surface force interference, flew antisubmarine patrols in the immediate area, and provided air-sea rescue services for carrier operations from D minus 1 day to the end of the campaign.

Army and Marine Corps troops landed on the western shores (1 Apr) against light opposition, established a firm beachhead, and captured Yontan airfield the same day. Supporting shore based air moved in behind the landings led by the OY-1 spotting planes (3 Apr). As Ground opposition stiffened, Marine Corps elements of the Tactical Air Force began local air defense patrols (7 Apr) and shortly started their close air support mission. A Navy landplane patrol squadron joined forces ashore (22 Apr) and extended the range of seaplane search operations, and an Army fighter squadron began operations from Ie Shima (13 May).

Strong Japanese air opposition developed (6 Apr) in the first of a series of mass suicide attacks involving some 400 aircraft. In seven mass raids, interspersed with smaller scattered ones, during the critical period (6 Apr-28 May), the Japanese expended some 1,500 aircraft, principally against naval forces supporting the campaign. In the 3 month's struggle against the humanly guided missiles of the Kamikaze force, the U.S. Navy took the heaviest punishment in its history. Although Task Force 58 lost no ship during the campaign, eight heavy and one light carriers were hit: *Enterprise, Intrepid, Yorktown* (18 Mar),

Appendix I

The Evolution of Aircraft Carriers

Franklin, Wasp (19 Mar), *San Jacinto* (6 Apr), *Hancock* (7 Apr) *Enterprise, Essex,* (11 Apr) *Intrepid* (16 Apr) *Bunker Hill* (11 May), and the *Enterprise* (14 May). Three escort carriers of Task Force 52, *Wake Island* (3 Apr), *Sangamon* (4 May), and *Natoma Bay* (6 June), were hit.

Opposition from Japanese naval forces was brief and ineffective. A task force of one light cruiser, eight destroyers, and *Yamato*, the world's largest battleship, made what was to be the last sortie of the Japanese Navy and was decisively beaten by carrier aircraft in the Battle of the East China Sea (7 Apr), in which only four Japanese destroyers survived.

Carrier air support was on a larger and more extensive scale than any previous amphibious campaign. Fast and escort carrier planes flew over 40,000 action sorties, destroyed 2,516 enemy aircraft, and blasted enemy positions with 8,500 tons of bombs and 50,000 rockets. Marine Corps squadrons ashore destroyed another 506 Japanese aircraft and expended 1,800 tons of bombs and 15,865 rockets on close air support missions. Task Force 58's time on the line (18 Mar-10 June) was surpassed by the escort carriers (24 Mar-21 June), but of several records for continuous operations in an active combat area that were marked up by the carriers during the campaign, the most outstanding was logged by the Essex with 79 consecutive days.

21--The development of a rocket-powered surface-to-air guided missile, was initiated as the Bureau of Aeronautics awarded a contract for 100 experimental Larks to the Ranger Engine Division of Fairchild.

26--Commander Fleet Air Wing 1, based on *Hamlin*, arrived at Kerama Retto to direct the operations of patrol squadrons assigned to support the assault and capture of Okinawa.

APRIL

14--Commander Fleet Air Wing 10 arrived at Puerto Princessa, Palawan, to direct patrol plane operations against the shipping in the South China Sea and along the Indo-China coast.

23--PB4Y's of Patrol Bombing Squadron 109 launched two Bat glide bombs against the enemy shipping in Balikpapan Harbor, Borneo, in the first combat employment of the only automatic homing bomb to be used in World War II.

MAY

1--CVBG-74, the first large Carrier Air Group in the U.S. Navy, was established at NAAF Otis Field, Mass., for duty on *Midway*.

2--First Helicopter Rescue--Lieutenant August Kleisch, USCG, flying a HNS-1 helicopter rescued 11 Canadian airmen that were marooned in northern Labrador about 125 miles from Goose Bay.

Appendix I

The Evolution of Aircraft Carriers

4--Fleet Air Wing 18, Rear Admiral M. R. Greer commanding, was established at Guam to take over the operational responsibilities in the Marianas area, previously held by Fleet Air Wing 1.

8--V-E Day--The President proclaimed the end of the war in Europe.

9--The U-249, the first German submarine to surrender after the cessation of hostilities in Europe, raised the black surrender flag to a PB4Y of Fleet Air Wing 7 near the Scilly Islands off Lands End England.

10--In a crash program to counter the Japanese Baka (suicide) bomb, the Naval Aircraft Modification Unit was authorized to develop Little Joe, a ship-to-air guided missile powered with a standard JATO unit.

19--The Office of Research and Inventions was established in the Office of the Secretary of the Navy to coordinate, and from time to time to disseminate to all bureaus full information with respect to all naval research, experimental, test and developmental activities and to supervise and administer all Navy Department action relating to patents, inventions, trademarks, copyrights, royalty payments, and similar matters. By this order, the Naval Research Laboratory and the Special Devices Division of the Bureau of Aeronautics were transferred to the newly established office.

JUNE

5--Cognizant commands and offices were informed of plans, permitted by the cessation of hostilities in Europe, for the future employment of Atlantic patrol aviation which called for the disestablishing of four Wings and 23 Patrol, five Inshore Patrol, and seven Composite Squadrons, and for the redeployment of seven Patrol Squadrons to the Pacific.

10--After the close of hostilities in Europe, Fleet Air Wing 15 departed from Port Lyautey for Norfolk.

13--A ramjet engine produced power in supersonic flight in a test conducted by the Applied Physics Laboratory at Island Beach, N.J. The ramjet unit was launched by a booster of four 5-inch high velocity aircraft rockets and achieved a range of 11,000 yards, nearly double that of similarly launched, cold units.

15--Fleet Airship Wing 2 at Richmond, Fla., was disestablished.

15--Experimental Squadrons XVF-200 and XVJ-25 were established at Brunswick, Maine, to provide, under the direct operational control of CominCh, flight facilities for evaluating and testing tactics, procedure, and equipment for use in special defense tasks particularly those concerned with defense against the Kamikaze.

16--Naval Air Test Center, Patuxent River, was established under a commander responsible for aviation test functions formerly assigned to NAS Patuxent River.

Appendix I

The Evolution of Aircraft Carriers

20--Fifth Wake Raid--Three carriers of Task Group 12.4 (Rear Admiral R. E. Jennings) launched five strikes against enemy positions on Wake Island.

27--Fleet Air Wing 16 was disestablished at Recife, Brazil.

30 June-3 July--Landings at Balikpapan--Marine Corps and Navy squadrons, aboard three escort carriers of Task Group 78.4 (Rear Admiral W. D. Sample), provided close air support, local combat air patrol, and strikes on military installations, in support of landings by Australian troops (1 July) at Balikpapan, Borneo.

JULY

10 July-15 August--Carrier Operations Against Japan--Task Force 38 (Vice Admiral J. S. McCain), initially composed of 14 carriers and augmented by one other later in the period, operated against the Japanese homeland in a series of air strikes on airfields, war and merchant shipping, naval bases and military installations from Kyushu in the south to Hokkaido in the north. The force was a part of Third Fleet under Admiral W. F. Halsey, who was in overall command. Operations were supported by a replenishment group and an antisubmarine group, both with escort carriers in their complement, and were supplemented (after 16 July) by operations of British Carrier Task Force 37 (Vice Admiral H. B. Rawlings, RN) of four carriers and screen.

The attack began with heavy air strikes on airfields in the Tokyo plains area (10 July), shifted to airfields and shipping in the northern Honshu-Hokkaido area (14-15 July), and returned to Tokyo targets (17 July) and naval shipping at Yokosuka (18 July). The attack hit Inland Sea shipping in the Kure area and airfields on northern Kyushu (24 July), swept up the Sea to the Osaka area and to Nagoya (25 July), and then repeated the sweep (25, 28 and 30 July). After moving southward (1 Aug.) to evade a typhoon, the force moved northward to clear the Hiroshima area for the atomic bomb drop and hit the Honshu-Hokkaido area (9-10 Aug), and Tokyo (13 Aug). On 15 August at 0635, when Admiral Halsey sent a message to his forces announcing the end of hostilities and ordering the cessation of offensive air operations, the first carrier strike of the day had already hit Tokyo and the second was approaching the coastline as it was recalled.

In this final carrier action of World War II, carrier aircraft destroyed 1,223 enemy aircraft of which over 1,000 were on the ground, and sank 23 war and 48 merchant ships totaling 285,000 tons.

13--Captain R. S. Barnaby, commanding the Johnsville Naval Aircraft Modification Unit, reported that the LBD-1, or Gargoyle, air-to-surface missile, in a series of 14 test flights including two at service weight, had made five satisfactory runs, thereby demonstrating that it was potentially capable of carrying out its mission.

14--Fleet Air Wing 12 was disestablished at Key West, Fla.

Appendix I

The Evolution of Aircraft Carriers

14--Commander Fleet Air Wing 7 embarked in Albemarle at Avonmouth, England, for transfer of headquarters to the United States at Norfolk.

14--Commander Fleet Air Wing 1, on Norton Sound, set up his command base in Chimu Wan, Okinawa, and directed patrol plane operations over the East China Sea, the Yellow Sea, and the coastal waters of Japan from that location until the end of the war.

15--Fleet Airship Wing 4 at Recife, Brazil, was disestablished.

18--Sixth Wake Raid--The carrier *Wasp*, returning to action after battle repairs and overhaul at Puget Sound, launched air strikes against targets on Wake.

19--Fleet Air Wing 9 was disestablished at NAS New York.

20--Little Joe, a rocket-propelled surface-to-air missile, made two successful flights at Applied Physics Laboratory (Johns Hopkins University) test station at Island Beach, N.J.

20--Fleet Airborne Electronics Training Units (FAETU) were established in the Atlantic and Pacific Fleets to train airborne early warning crews in the theory, operation and maintenance of their equipment.

24--Marine Corps pilots, operating from the escort carrier *Vella Gulf*, attacked Japanese positions on Pagan Island in the Marianas, and 2 days later hit Rota in the same island group.

28--Fleet Air Wing 15 was disestablished at Norfolk.

AUGUST

1--Seventh Wake Raid--Task Group 12.3, composed of one carrier, one battleship and destroyer screen, bombed and bombarded Wake.

4--Fleet Air Wing 7 was disestablished at Norfolk.

6--Eighth Wake Raid--The carrier Intrepid, while en route from Pearl Harbor to join forces off Japan, bombed buildings and gun positions on Wake Island.

6--Escort carriers from TG 95.3 (Rear Admiral C. T. Durgin), covering a cruiser force operating in the East China Sea, launched strikes on shipping in the harbor at Tinghai, China.

14--Japan accepted the terms of unconditional surrender and on the same day, which was the 15th in the Western Pacific, hostilities ceased.

21--The Asiatic Wing, Naval Air Transport Service, was established at NAS Oakland, Captain E. F. Luethi in command, to operate and maintain air transport support of establishments and units in the Western Pacific and Asiatic theaters. Early in September, Wing headquarters was established on Samar in the Philippines, and on 15 November transferred to NAB Agana, Guam.

SEPTEMBER

Appendix I

The Evolution of Aircraft Carriers

2--The formal surrender of Japan, on board *Missouri* in Tokyo Bay, marked V-J Day and the end of World War II.

10--*Midway*, first of the 45,000 ton class aircraft carriers, was placed in commission at Newport News, Va., with Captain Joseph F. Bolger in command.

Appendix I

The Evolution of Aircraft Carriers

OCTOBER

3--As the initial attempt to establish an earth satellite program, the Bureau of Aeronautics established a committee to evaluate the feasibility of space rocketry.

10--The Office of Chief of Naval Operations was reorganized and four new Deputy Chiefs were set up for Personnel, Administration, Operations and Logistics on the same level as the existing Deputy Chief of Naval Operations (Air). The reorganization, which was by direction of the Secretary and in accord with Executive Order, abolished Commander in Chief, U.S. Fleet, and transferred command of the operating forces to the Chief of Naval Operations.

17--A type designation letter K for pilotless aircraft was added to the basic designation system, replacing the previous Class designation VK. Classes A, G and S within the type were assigned for pilotless aircraft intended for attack against aircraft, ground targets, and ships respectively.

29--The Committee to Evaluate the Feasibility of Space Rocketry recommended that detailed studies be made to determine the feasibility of an Earth Satellite Vehicle. This led the Bureau of Aeronautics to issue contracts to one university and three companies for theoretical study, and preliminary design of a launch vehicle and for determining by actual test the specific impulse of high energy fuels including liquid hydrogen-oxygen.

NOVEMBER

1--The Naval Air Training Command was reorganized with headquarters at NAS Pensacola, and the following subordinate commands: Naval Air Advanced Training, Naval Air Basic Training, Naval Air Technical Training, and a newly formed Naval Air Reserve Training. By this change the titles Naval Air Operational Training and Naval Air Intermediate Training ceased to exist and the facilities of the former Naval Air Primary Training Command were incorporated into Basic Training or absorbed by the Reserve Program.

5--Ensign Jake C. West, with VF-41 embarked on Wake Island for carrier qualifications with the FR-1, lost power on the forward radial engine of his FR-1 shortly after take-off, forcing him to start his aft jet engine. He returned to the ship and made a successful landing, the first jet landing aboard a carrier.

29--The Special Weapons Test and Tactical Evaluation Unit was redesignated Pilotless Aircraft Unit and in the next month was transferred to MCAS Mojave, and directed to operate detachments at NAF Point Mugu as necessary.

DECEMBER

1--Fleet Air Wing 6 was disestablished at NAS Whidbey Island.

28--The President directed that the Coast Guard be transferred from the Navy and returned to the jurisdiction of the Treasury Department.

Appendix II

Attack Carrier Designations and Names

Appendix II

Carrier Designations and Names

Attack Carriers (CV, CVA, CVB, CVL, CVAN, CVN)

The CVB and CVL designations were established within the original CV designation on 15 July 1943. CVA replaced CV and CVB on 1 October 1952; CVL went out of use on 15 May 1959. CV and CVN replaced CVA and CVAN on 30 June 1975 to designate the multi-mission character of aircraft carriers after the decommissioning of the last CVS in 1974.

During World War II (7 December 1941 to 2 September 1945) the Navy operated 110 carriers (includes those designated CV, CVE and CVL). It also commissioned 102 new carriers (including those designated CV, CVE and CVL) during the above mentioned time frame. The Navy also operated two training carriers during World War II with the designation IX. They were USS Wolverine (IX 64) and USS Sable (IX 81).

Original Classes

Langley Class: 1 ship (CV 1)
Lexington Class: 2 ships (CV 2 and 3)
Ranger Class: 1 ship (CV 4)
Yorktown Class: 2 ships (CV 5 and 6)
Wasp Class: 1 ship (CV 7)
Hornet Class: 1 ship (CV 8)
Essex Class: 24 ships; CV 9 through 21, 31 through 35, 37 through 40, 45, and 47.

(Long-Hull Essex Class or Ticonderoga Class): Of these numbers - 14, 15, 19, 21, 32-34, 36-40, 45 and 47 are sometimes referred to as "Long-Hull" Essex class or Ticonderoga Class

Independence Class: 9 ships, CVL 22 through 30.
Midway Class: 3 ships, CVB 41 through 43.
Saipan Class: 2 ships, CVL 48 and 49.
Enterprise Class: 1 ship, CVAN-65.
Forrestal Class: 4 ships, CVA 59 through 62.
Kitty Hawk Class: 4 ships, CVA 63, 64, 66 and 67.
Nimitz Class: 9 ships, CVN 68 through 76.

Carrier Listing for CV, CVA, CVB, CVAN, CVN, and CVL

Hull No. and Name: 1 Langley
Date of Commission: 20 Mar 1922; Decommission or Loss***: 27 Feb 1942
Designations: CV 1 (21 Apr 1937); AV-3 (8 May 1942)
Conversion Project*/Date completed:
Comments: Lost, enemy action

Hull No. and Name: 2 Lexington
Date of Commission: 14 Dec 1927; Decommission or Loss***: 8 May 1942
Designations: CV 2
Conversion Project*/Date completed:
Comments: Lost, enemy action

Appendix II

The Evolution of Aircraft Carriers

Hull No. and Name: 3 Saratoga
Date of Commission: 16 Nov 1927; Decommission or Loss***:
Designations: CV 3
Conversion Project*/Date completed:
Comments: Expended, Operation Crossroads, 26 Jul 1946

Hull No. and Name: 4 Ranger
Date of Commission: 4 Jun 1934; Decommission or Loss***: 18 Oct 1946
Designations: CV 4
Conversion Project*/Date completed:
Comments: Sold for scrap 31 Jan 1947

Hull No. and Name: 5 Yorktown
Date of Commission: 30 Sep 1937; Decommission or Loss***: 7 Jun 1942
Designations: CV 5
Conversion Project*/Date completed:
Comments: Lost, enemy action

Hull No. and Name: 6 Enterprise
Date of Commission: 12 May 1938; Decommission or Loss***: 17 Feb 1947
Designations: CV 6 ; CVA 6 (1 Oct 1952); CVS 6 (8 Aug 1953)
Conversion Project*/Date completed:
Comments: Sold, 1 Jul 1958

Hull No. and Name: 7 Wasp
Date of Commission: 25 Apr 1940; Decommission or Loss***: 15 Sep 1942
Designations: CV 7
Conversion Project*/Date completed:
Comments: Lost, enemy action

Hull No. and Name: 8 Hornet
Date of Commission: 20 Oct 1941; Decommission or Loss***: 26 Oct 1942
Designations: CV 8
Conversion Project*/Date completed:
Comments: Lost, enemy action

Hull No. and Name: 9 Essex
Date of Commission: 31 Dec 1942; Decommission or Loss***: 30 Jun 1969
Designations: CV 9; CVA 9 (1 Oct 1952); CVS 9 (8 Mar 1960)
Conversion Project*/Date completed: 27A (Feb 1951); 125 (Mar 1956)
Comments: Stricken 1 Jun 1973

Hull No. and Name: 10 Yorktown
Date of Commission: 15 Apr 1943; Decommission or Loss***: 27 Jun 1970
Designations: CV 10; CVA 10 (1 Oct 1952); CVS 10 (1 Sep 1957)
Conversion Project*/Date completed: 27A (Jan 1953); 125 (Oct 1955)
Comments: Stricken 1 Jun 1973, Floating museum at Charleston, S.C. ,13 Nov 1975

Hull No. and Name: 11 Intrepid
Date of Commission: 16 Aug 1943; Decommission or Loss***: 15 Mar 1974
Designations: CV 11; CVA 11 (1 Oct 1952); CVS 11 (31 Mar 1962)
Conversion Project*/Date completed: 27C (Jun 1954); 27C (Apr 1957)
Comments: Floating museum, New York City

Appendix II

The Evolution of Aircraft Carriers

Hull No. and Name: 12 Hornet
Date of Commission: 29 Nov 1943; Decommission or Loss***: 26 May 1970
Designations: CV 12; CVA 12 (1 Oct 1952); CVS 12 (27 Jun 1958)
Conversion Project*/Date completed: 27A (Oct 1953); 125 (Aug 1956)
Comments: Stricken 1989

Hull No. and Name: 13 Franklin
Date of Commission: 31 Jan 1944; Decommission or Loss***: 17 Feb 1947
Designations: CV 13; CVA 13 (1 Oct 1952); CVS 13 (8 Aug 1953)
Conversion Project*/Date completed:
Comments: Stricken 10 Oct 1964

Hull No. and Name: 14 Ticonderoga
Date of Commission: 8 May 1944; Decommission or Loss***: 1 Sep 1973
Designations: CV 14; CVA 14 (1 Oct 1952); CVS 14 (21 Oct 1969)
Conversion Project*/Date completed: 27C (Dec 1954); 27C (Mar 1957)
Comments: Stricken 16 Nov 1973

Hull No. and Name: 15 Randolph
Date of Commission: 9 Oct 1944; Decommission or Loss***: 13 Feb 1969
Designations: CV 15; CVA 15 (1 Oct 1952); CVS 15 (31 Mar 1959)
Conversion Project*/Date completed: 27A (Jul 1953); 125 (Feb 1956)
Comments: Stricken 1 Jun 1973

Hull No. and Name: 16 Lexington
Date of Commission: 17 Feb 1943; Decommission or Loss***: 8 Nov 1991
Designations: CV 16; CVA 16 (1 Oct 1952); CVS 16 (1 Oct 1962); CVT 16 (1 Jan 1969); AVT 16 (1 Jul 1978)
Conversion Project*/Date completed: 27C (Sep 1955)
Comments: Stricken 30 Nov 1991

Hull No. and Name: 17 Bunker Hill
Date of Commission: 25 May 1943; Decommission or Loss***: 9 Jul 1947
Designations: CV 17; CVA 17 (1 Oct 1952); CVS 17 (8 Aug 1953)
Conversion Project*/Date completed:
Comments: Stricken 1 Nov 1966, retained as moored electronic test ship San Diego until Nov 1972; Scrapped 1973

Hull No. and Name: 18 Wasp
Date of Commission: 24 Nov 1943; Decommission or Loss***: 1 Jul 1972
Designations: CV 18; CVA 18 (Oct 1952); CVS 18 (1 Nov 1956)
Conversion Project*/Date completed: 27A (Sep 1951); 125 (Dec 1955)
Comments: Sold for scrap 21 May 1973

Hull No. and Name: 19 Hancock
Date of Commission: 15 Apr 1944; Decommission or Loss***: 30 Jan 1976
Designations: CV 19; CVA 19 (1 Oct 1952); CV 19 (30 Jun 1975)
Conversion Project*/Date completed: 27C (Mar 1954); 17C (Nov 1956)
Comments: Stricken 31 Jan 1976

Hull No. and Name: 20 Bennington
Date of Commission: 6 Aug 1944; Decommission or Loss***: 15 Jan 1970
Designations: CV 20; CVA 20 (1 Oct 1952); CVS 20 (30 Jun 1959)
Conversion Project*/Date completed: 27A (Nov 1952); 125 (Apr 1955)
Comments: Stricken 1989

Appendix II

The Evolution of Aircraft Carriers

Hull No. and Name: 21 Boxer
Date of Commission: 16 Apr 1945; Decommission or Loss***: 1 Dec 1969
Designations: CV 21; CVA 21 (1 Oct 1952); CVS 21 (1 Feb 1956); LPH 4 (30 Jan 1959)
Conversion Project*/Date completed:
Comments: Stricken 1 Dec 1969.

Hull No. and Name: 22 Independence
Date of Commission: 14 Jan 1943; Decommission or Loss***: 28 Aug 1946
Designations: CVL 22
Conversion Project*/Date completed:
Comments: Sunk in weapons test 29 Jan 1951

Hull No. and Name: 23 Princeton
Date of Commission: 25 Feb 1943; Decommission or Loss***: 24 Oct 1944
Designations: CVL 23
Conversion Project*/Date completed:
Comments: Lost, enemy action.

Hull No. and Name: 24 Belleau Wood
Date of Commission: 31 Mar 1943; Decommission or Loss***: 13 Jan 1947
Designations: CVL 24
Conversion Project*/Date completed:
Comments: To France 1953-1960; Stricken 1 Oct 1960

Hull No. and Name: 25 Cowpens
Date of Commission: 28 May 1943; Decommission or Loss***: 13 Jan 1947
Designations: CVL 25; AVT 1 (15 May 1959)
Conversion Project*/Date completed:
Comments: Stricken 1 Nov 1959

Hull No. and Name: 26 Monterey
Date of Commission: 17 Jun 1943; Decommission or Loss***: 16 Jan 1956
Designations: CVL 26; AVT 2 (15 May 1959)
Conversion Project*/Date completed:
Comments: Stricken 1 Jun 1970

Hull No. and Name: 27 Langley
Date of Commission: 31 Aug 1943; Decommission or Loss***: 11 Feb 1947
Designations: CVL 27
Conversion Project*/Date completed:
Comments: Transferred to France 1951- 1963. Sold 19 Feb 1964

Hull No. and Name: 28 Cabot
Date of Commission: 24 Jul 1943; Decommission or Loss***: 30 Aug 1967
Designations: CVL 28 AVT 3 15 May 1959
Conversion Project*/Date completed:
Comments: Transferred to Spain on 21 Jan 1955, returned to private U.S. organization in 1989

Hull No. and Name: 29 Bataan
Date of Commission: 17 Nov 1943; Decommission or Loss***: 9 Apr 1954
Designations: CVL 29; AVT 4 (15 May 1959)
Conversion Project*/Date completed:
Comments: Stricken 1 Sep 1959

Appendix II

The Evolution of Aircraft Carriers

Hull No. and Name: 30 San Jacinto
Date of Commission: 15 Dec 1943; Decommission or Loss***: 1 Mar 1947
Designations: CVL 30; AVT 5 (15 May 1959)
Conversion Project*/Date completed:
Comments: Stricken 1 Jun 1970

Hull No. and Name: 31 Bon Homme Richard
Date of Commission: 26 Nov 1944; Decommission or Loss***: 2 Jul 1971
Designations: CV 31; CVA 31 (1 Oct 1952)
Conversion Project*/Date completed:
Comments: Stricken 1989

Hull No. and Name: 32 Leyte
Date of Commission: 11 Apr 1946; Decommission or Loss***: 15 May 1959
Designations: CV 32; CVA 32 (1 Oct 1952); CVS 32 (8 Aug 1953); AVT 10 (15 May 1959)
Conversion Project*/Date completed:
Comments: Stricken 1 Jun 1969

Hull No. and Name: 33 Kearsarge
Date of Commission: 2 Mar 1946; Decommission or Loss***: 15 Jan 1970
Designations: CV 33; CVA 33 (1 Oct 1952); CVS 33 (1 Oct 1958)
Conversion Project*/Date completed: 27A (Mar 1952); 125 (Jan 1957)
Comments: Stricken 1 May 1973

Hull No. and Name: 34 Oriskany
Date of Commission: 25 Sep 1950; Decommission or Loss***: 20 Sep 1976
Designations: CV 34; CVA 34 (1 Oct 1952); CV 34 (30 Jun 1975)
Conversion Project*/Date completed: 27A (Oct 1950); 125 (May 1959)
Comments: Stricken 1989

Hull No. and Name: 36 Antietam
Date of Commission: 28 Jan 1945; Decommission or Loss***: 8 May 1963
Designations: CV 36; CVA 36 (1 Oct 1952); CVS 36 (8 Aug 1953)
Conversion Project*/Date completed: **
Comments: Stricken 1 May 1973

Hull No. and Name: 37 Princeton
Date of Commission: 18 Nov 1945; Decommission or Loss***: 30 Jan 1970
Designations: CV 37; CVA 37 (1 Oct 1952); CVS 37 (1 Jan 1954); LPH 5 (2 Mar 1959)
Conversion Project*/Date completed:
Comments: Stricken 30 Jan 1970

Hull No. and Name: 38 Shangri-La
Date of Commission: 15 Sep 1944; Decommission or Loss***: 30 Jul 1971
Designations: CV 38; CVA 38 (1 Oct 1952); CVS 38 (30 Jun 1969)
Conversion Project*/Date completed: 27C (Feb 1955)
Comments: Stricken 15 Jul 1982

Hull No. and Name: 39 Lake Champlain
Date of Commission: 3 Jun 1945; Decommission or Loss***: 2 May 1966
Designations: CV 39; CVA 39 (1 Oct 1952); CVS 39 (1 Aug 1957)
Conversion Project*/Date completed: 27A (Sep 1952)
Comments: Stricken 1 Dec 1969

Appendix II

The Evolution of Aircraft Carriers

Hull No. and Name: 40 Tarawa
Date of Commission: 8 Dec 1945; Decommission or Loss***: 13 May 1960
Designations: CV 40; CVA 40 (1 Oct 1952); CVS 40 (10 Jan 1955); AVT 12 (17 Apr 1961)
Conversion Project*/Date completed: 27A (Sep 1952)
Comments: Stricken 1 Jun 1967

Hull No. and Name: 41 Midway
Date of Commission: 10 Sep 1945; Decommission or Loss***: 11 Apr 1992
Designations: CVB 41; CVA 41 (1 Oct 1952); CV 41 (30 Jun 1975)
Conversion Project*/Date completed: 110 (Nov 1957)
Comments: In reserve

Hull No. and Name: 42 Franklin D. Roosevelt
Date of Commission: 27 Oct 1945; Decommission or Loss***: 1 Oct 1977
Designations: CVB 42; CVA 42 (1 Oct 1952); CV 42 (30 Jun 1975)
Conversion Project*/Date completed: 110 (Jun 1956)
Comments: Stricken 30 Sep 1977

Hull No. and Name: 43 Coral Sea
Date of Commission: 1 Oct 1947; Decommission or Loss***: 26 Apr 1990
Designations: CVB 43; CVA 43 (1 Oct 1952); CV 43 (30 Jun 1975)
Conversion Project*/Date completed: 110A (Jan 1960)
Comments:

Hull No. and Name: 45 Valley Forge
Date of Commission: 3 Nov 1946; Decommission or Loss***: 15 Jan 1970
Designations: CV 45; CVA 45 (1 Oct 1952); CVS 45 (1 Jan 1954); LPH 8 (1 Jul 1961)
Conversion Project*/Date completed:
Comments: Stricken 15 Jan 1970

Hull No. and Name: 47 Philippine Sea
Date of Commission: 11 May 1946; Decommission or Loss***: 28 Dec 1958
Designations: CV 47; CVA 47 (1 Oct 1952); CVS 47 (15 Nov 1955); AVT 11 (15 May 1959)
Conversion Project*/Date completed:
Comments: Stricken 1 Dec 1969

Hull No. and Name: 48 Saipan
Date of Commission: 14 Jul 1946; Decommission or Loss***: 14 Jan 1970
Designations: CVL 48; AVT 6 (15 May 1959); AGMR 2 (8 Apr 1965)
Conversion Project*/Date completed:
Comments:

Hull No. and Name: 49 Wright
Date of Commission: 9 Feb 1947; Decommission or Loss***: 22 May 1970
Designations: CVL 49; AVT 7 (15 May 1959); CC 2 (11 May 1963)
Conversion Project*/Date completed:
Comments:

Hull No. and Name: 59 Forrestal
Date of Commission: 1 Oct 1955; Decommission or Loss***: 30 Sep 1993
Designations: CVA 59; CV 59 (30 Jun 1975); AVT 59 (4 Feb 1992)
Conversion Project*/Date completed:
Comments: Stricken 11 Sep 1993

Appendix II

The Evolution of Aircraft Carriers

Hull No. and Name: 60 Saratoga
Date of Commission: 14 Apr 1956; Decommission or Loss***: 20 Aug 1994
Designations: CVA 60; CV 60 (30 Jun 1972)
Conversion Project*/Date completed:
Comments: Stricken 30 Sep 1994

Hull No. and Name: 61 Ranger
Date of Commission: 10 Aug 1957; Decommission or Loss***: 10 Jul 1993
Designations: CVA 61; CV 61 (30 Jun 1975)
Conversion Project*/Date completed:
Comments: Inactive in Reserve

Hull No. and Name: 62 Independence
Date of Commission: 10 Jan 1959; Decommission or Loss***:
Designations: CVA 62; CV 62 (28 Feb 1973)
Conversion Project*/Date completed:
Comments: Inactive.

Hull No. and Name: 63 Kitty Hawk
Date of Commission: 29 Apr 1961; Decommission or Loss***:
Designations: CVA 63; CV 63 (29 Apr 1973)
Conversion Project*/Date completed:
Comments: Inactive

Hull No. and Name: 64 Constellation
Date of Commission: 27 Oct 1961; Decommission or Loss***:
Designations: CVA 64; CV 64 (30 Jun 1975)
Conversion Project*/Date completed:
Comments: Inactive

Hull No. and Name: 65 Enterprise
Date of Commission: 25 Nov 1961; Decommission or Loss***:
Designations: CVAN 65; CVN 65 (30 Jun 1975)
Conversion Project*/Date completed:
Comments: Inactive as of December 2012

Hull No. and Name: 66 America
Date of Commission: 23 Jan 1965; Decommission or Loss***: 5 Sep 1996
Designations: CVA 66; CV 66 (30 Jun 1975)
Conversion Project*/Date completed:
Comments: Inactive in Reserve

Hull No. and Name: 67 John F. Kennedy
Date of Commission: 7 Sep 1968; Decommission or Loss***:
Designations: CVA 67; CV 67 (29 Apr 1973)
Conversion Project*/Date completed:
Comments: Inactive

Hull No. and Name: 68 Nimitz
Date of Commission: 3 May 1975; Decommission or Loss***:
Designations: CVAN 68; CVN 68 (30 Jun 1975)
Conversion Project*/Date completed:
Comments: Active

Appendix II

The Evolution of Aircraft Carriers

Hull No. and Name: 69 Dwight D. Eisenhower
Date of Commission: 18 Oct 1977; Decommission or Loss***:
Designations: CVN 69
Conversion Project*/Date completed:
Comments: Active

Hull No. and Name: 70 Carl Vinson
Date of Commission: 13 Mar 1982; Decommission or Loss***:
Designations: CVN 70
Conversion Project*/Date completed:
Comments: Active

Hull No. and Name: 71 Theodore Roosevelt
Date of Commission: 25 Oct 1986; Decommission or Loss***:
Designations: CVN 71
Conversion Project*/Date completed:
Comments: Active

Hull No. and Name: 72 Abraham Lincoln
Date of Commission: 11 Nov 1989; Decommission or Loss***:
Designations: CVN 72
Conversion Project*/Date completed:
Comments: Active

Hull No. and Name: 73 George Washington
Date of Commission: 4 Jul 1992; Decommission or Loss***:
Designations: CVN 73
Conversion Project*/Date completed:
Comments: Active

Hull No. and Name: 74 John C. Stennis
Date of Commission: 9 Dec 1995; Decommission or Loss***:
Designations: CVN 74
Conversion Project*/Date completed:
Comments: Active

Hull No. and Name: 75 Harry S. Truman
Date of Commission: ; Decommission or Loss***:
Designations: CVN 75
Conversion Project*/Date completed:
Comments: Keel laid 29 Nov 1993

Hull No. and Name: 76 Ronald Reagan
Date of Commission: 7 Dec 2003; Decommission or Loss***:
Designations: CVN 76
Conversion Project*/Date completed:
Comments: Active

Hull No. and Name: 77 George W Bush
Date of Commission: 10 Jan 2009; Decommission or Loss***:
Designations: CVN 76
Conversion Project*/Date completed:
Comments: Active

Appendix II

The Evolution of Aircraft Carriers

Notations for Appendix II

* Projects 27A and the first 27Cs are axial deck modernizations; all others are angled deck conversions. For more detail, see chronology entries for 4 Jun 1947, 1 Feb 1952, 2 Sep 1953 and 27 May 1954.

** Experimental angle deck installation completed Dec 1952.

*** There were a number of carriers that were decommissioned and then re-commissioned for further service. Only the final decommissioning date is listed for these carriers. Several carriers were also placed out of commission during major renovations or yard periods.

Note 1: Construction of hull numbers omitted above were either terminated or cancelled. Numbers 35, 46, and 50-55 were scheduled for Essex class; 44, 56 and 57 for *Midway* class. Number 58 was *United States*. Note 2: The contracts originally set for CV 59 and 60 (*Forrestal* and *Saratoga*) did not include an angled deck in their designs. In 1953 the Navy redesigned the flight deck plans for the *Forrestal* and incorporated an angled landing deck. These changes were also made to the designs for the *Saratoga*. The contract for *Forrestal* was awarded in 1951 and for *Saratoga* in 1952. The contract for *Ranger* and *Independence* (CV 61 and 62) were not awarded until 1954. Therefore, the original contract designs for the *Ranger* and *Independence* would have included an angled deck. Technically speaking, *Ranger* (CVA 61) was the first carrier designed and built as an angled deck carrier.

Appendix III

Escort Carrier Designations and Names

Appendix III

Carrier Designations and Names

Escort Carriers (AVG, ACV and CVE)

The original escort carrier designation AVG (Aircraft Escort Vessel) was first assigned on 31 March 1941. The classification was changed to ACV (Auxiliary Aircraft Carrier) on 20 August 1942 and to CVE (Escort Carrier) on 15 July 1943.

The CVE designation went out of use when the remaining escort carriers were reclassified AKV (Aircraft Ferry) on 7 May 1959.

Classes:

Long Island Class:	1 ship, hull number 1
Charger Class:	1 ship, hull number 30 (originaly built for Royal Navy)
Bogue Class:	11 ships, hull numbers 9, 11-13, 16, 18, 20, 21, 23, 25 and 31
Sangamon Class:	4 ships, hull numbers 26-29
Casablanca Class:	50 ships, hull numbers 55-104
Commencement Bay Class:	19 ships, hull numbers 105-123

Hull numbers not listed below are accounted for as follows:

2-5 not assigned; 6, 7, 8, 10, 14, 15, 17, 19, 22, 24, 32-54 transferred to the UK; 124-139 cancelled.

Carrier Listing for CVE Designations

Hull No. and Name: 1 Long Island
Date of Commission: 2 Jun 1941; Decommission or Loss***: 20 Mar 1946
Designations:
Comments: Stricken 12 Apr 1946

Hull No. and Name: 9 Bogue
Date of Commission: 26 Sep 1942; Decommission or Loss***: 30 Nov 1946
Designations: CVHE 9 (12 Jun 1955)
Comments: Stricken 1 Mar 1959

Hull No. and Name: 11 Card
Date of Commission: 8 Nov 1942; Decommission or Loss***: 13 May 1946
Designations: CVHE 11 (12 Jun 1955); CVU 11 (1 Jun 1959); AKV 40 (7 May 1959)
Comments: Stricken 15 Sep 1970

Hull No. and Name: 12 Copahee
Date of Commission: 15 Jun 1942; Decommission or Loss***: 5 Jul 1946
Designations: CVHE 12 (12 Jun 1955)
Comments: Stricken 1 Mar 1959

Hull No. and Name: 13 Core
Date of Commission: 10 Dec 1942; Decommission or Loss***: 4 Oct 1946
Designations: CVHE 13 (12 Jun 1955); CVU 13 (1 Jul 1958); AKV 41 (7 May 1959)
Comments: Stricken 15 Sep 1970

Appendix III

The Evolution of Aircraft Carriers

Hull No. and Name: 16 Nassau
Date of Commission: 20 Aug 1942; Decommission or Loss***: 28 Oct 1946
Designations: CVHE 16 (12 Jun 1955)
Comments: Stricken 1 Mar 1959

Hull No. and Name: 18 Altamaha
Date of Commission: 15 Sep 1942; Decommission or Loss***: 27 Sep 1946
Designations: CVHE 18 (12 Jun 1955)
Comments: Stricken 1 Mar 1959

Hull No. and Name: 20 Barnes
Date of Commission: 20 Feb 1943; Decommission or Loss***: 29 Aug 1946
Designations: CVHE 20 (12 Jun 1955)
Comments: Stricken 1 Mar 1959

Hull No. and Name: 21 Block Island
Date of Commission: 8 Mar 1943; Decommission or Loss***: 29 May 1944
Designations:
Comments: Lost to enemy action

Hull No. and Name: 23 Breton
Date of Commission: 12 Apr 1943; Decommission or Loss***: 30 Aug 1946
Designations: CVHE 23 (12 Jun 1955); CVU 23 (1 Jul 1958); AKV 42 (7 May 1959)
Comments: Stricken 6 Aug 1971

Hull No. and Name: 25 Croatan
Date of Commission: 28 Apr 1943; Decommission or Loss***: 20 May 1946
Designations: CVHE 25 (12 Jun 1955); CVU 25 (1 Jul 1958); AKV 43 (7 May 1959)
Comments: Stricken 15 Sep 1970

Hull No. and Name: 26 Sangamon
Date of Commission: 25 Aug 1942; Decommission or Loss***: 24 Oct 1945
Designations: AO 28 23 Oct 1940 AVG 26 14 Feb 1942
Comments: Stricken 1 Nov 1945. Sangamon was a fleet oiler (AO 28) before being converted to an escort carrier.

Hull No. and Name: 27 Suwannee
Date of Commission: 24 Sep 1942; Decommission or Loss***: 8 Jan 1947
Designations: AO 33 (16 Jul 1941); AVG 27 (14 Feb 1942); CVHE 27 (12 Jun 1955)
Comments: Stricken 1 Mar 1959. Suwannee was a fleet oiler (AO 33) before being converted to an escort carrier.

Hull No. and Name: 28 Chenango
Date of Commission: 19 Sep 1942; Decommission or Loss***: 14 Aug 1946
Designations: AO 31 (20 Jun 1941); ACV 28 (19 Sep 1942); CVHE 28 (12 Jun 1955)
Comments: Stricken 1 Mar 1959. Chenango was a fleet oiler (AO 31) before being converted to an escort carrier

Hull No. and Name: 29 Santee
Date of Commission: 24 Aug 1942; Decommission or Loss***: 21 Oct 1946
Designations: AO 29 (30 Oct 1940); ACV 29 (24 Aug 1942); CVHE 29 (12 Jun 1955)
Comments: Stricken 1 Mar 1959. Santee was a fleet oiler (AO 29) before being converted to an escort carrier

Appendix III

The Evolution of Aircraft Carriers

Hull No. and Name: 30 Charger
Date of Commission: 3 Mar 1942; Decommission or Loss***: 15 Mar 1946
Designations:
Comments: Stricken 29 Mar 1946

Hull No. and Name: 31 Prince William
Date of Commission: 9 Apr 1943; Decommission or Loss***: 29 Aug 1946
Designations: CVHE 31 12 Jun 1955
Comments: Stricken 1 Mar 1959

Hull No. and Name: 55 Casablanca
Date of Commission: 8 Jul 1943; Decommission or Loss***: 10 Jun 1946
Designations:
Comments: Sold 23 Apr 1947

Hull No. and Name: 56 Liscome Bay
Date of Commission: 7 Aug 1943; Decommission or Loss***: 24 Nov 1943
Designations:
Comments: Lost to enemy action

Hull No. and Name: 57 Anzio (ex-Coral Sea)
Date of Commission: 27 Aug 1943; Decommission or Loss***: 5 Aug 1946
Designations: CVHE 57 12 Jun 1955
Comments: Stricken 1 Mar 1959

Hull No. and Name: 58 Corregidor
Date of Commission: 31 Aug 1943; Decommission or Loss***: 4 Sep 1958
Designations: CVU 58 12 Jun 1955
Comments: Stricken 1 Oct 1958

Hull No. and Name: 59 Mission Bay
Date of Commission: 13 Sep 1943; Decommission or Loss***: 21 Feb 1947
Designations: CVU 59 12 Jun 1955
Comments: Stricken 1 Sep 1958

Hull No. and Name: 60 Guadalcanal
Date of Commission: 25 Sep 1943; Decommission or Loss***: 15 Jul 1946
Designations: CVU 60 12 Jun 1955
Comments: Stricken 27 May 1958

Hull No. and Name: 61 Manila Bay
Date of Commission: 5 Oct 1943; Decommission or Loss***: 31 Jul 1946
Designations: CVU 61 12 Jun 1955
Comments: Stricken 27 May 1958

Hull No. and Name: 62 Natoma Bay
Date of Commission: 14 Oct 1943; Decommission or Loss***: 20 May 1946
Designations: CVU 62 12 Jun 1955
Comments: Stricken 27 May 1958

Hull No. and Name: 63 St. Lo (ex-Midway)
Date of Commission: 23 Oct 1943; Decommission or Loss***: 25 Oct 1944
Designations:
Comments: Lost to enemy action. Commissioned on 23 Oct 1943 as 25 Oct 1944 USS Midway (CVE 63) and renamed St. Lo (CVE 63) on 10 Oct 1944

Appendix III

The Evolution of Aircraft Carriers

Hull No. and Name: 64 Tripoli
Date of Commission: 31 Oct 1943; Decommission or Loss***: 25 Nov 1958
Designations: CVU 64 12 Jun 1955
Comments: Stricken 1 Feb 1959

Hull No. and Name: 65 Wake Island
Date of Commission: 7 Nov 1943; Decommission or Loss***: 5 Apr 1946
Designations:
Comments: Stricken 17 Apr 1946

Hull No. and Name: 66 White Plains
Date of Commission: 15 Nov 1943; Decommission or Loss***: 10 Jul 1946
Designations: CVU 66 12 Jun 1955
Comments: Stricken 1 Jul 1958

Hull No. and Name: 67 Solomons
Date of Commission: 21 Nov 1943; Decommission or Loss***: 15 May 1946
Designations:
Comments: Stricken 5 Jun 1946. Launched as USS Nassuk Bay (CVE 67) on 6 Oct 1943 and renamed Solomons (CVE 67) in Nov 1943

Hull No. and Name: 68 Kalinin Bay
Date of Commission: 27 Nov 1943; Decommission or Loss***: 15 May 1946
Designations:
Comments: Stricken 5 Jun 1946

Hull No. and Name: 69 Kasaan Bay
Date of Commission: 4 Dec 1943; Decommission or Loss***: 6 Jul 1946
Designations: CVHE 69 12 Jun 1955
Comments: Stricken 1 Mar 1959

Hull No. and Name: 70 Fanshaw Bay
Date of Commission: 9 Dec 1943; Decommission or Loss***: 14 Aug 1946
Designations: CVHE 70 12 Jun 1955
Comments: Stricken 1 Mar 1959

Hull No. and Name: 71 Kitkun Bay
Date of Commission: 15 Dec 1943; Decommission or Loss***: 19 Apr 1946
Designations:
Comments: Sold 18 Nov 1946

Hull No. and Name: 72 Tulagi
Date of Commission: 21 Dec 1943; Decommission or Loss***: 30 Apr 1946
Designations:
Comments: Stricken 8 May 1946

Hull No. and Name: 73 Gambier Bay
Date of Commission: 28 Dec 1943; Decommission or Loss***: 25 Oct 1944
Designations:
Comments: Lost to enemy action

Hull No. and Name: 74 Nehenta Bay
Date of Commission: 3 Jan 1944; Decommission or Loss***: 15 May 1946
Designations: CVU 74 (12 Jun 1955); AKV 24 (7 May 1959)
Comments: Stricken 1 Aug 1959

Appendix III

The Evolution of Aircraft Carriers

Hull No. and Name: 75 Hoggatt Bay
Date of Commission: 11 Jan 1944; Decommission or Loss***: 20 Jul 1946
Designations: CVHE 75 (12 Jun 1955); AKV 25 (7 May 1959)
Comments: Stricken 1 Aug 1959

Hull No. and Name: 76 Kadashan Bay
Date of Commission: 18 Jan 1944; Decommission or Loss***: 14 Jun 1946
Designations: CVU 76 (12 Jun 1955); AKV 26 (7 May 1959)
Comments: Stricken 1 Aug 1959

Hull No. and Name: 77 Marcus Island
Date of Commission: 26 Jan 1944; Decommission or Loss***: 12 Dec 1946
Designations: CVHE 77 (12 Jun 1955); AKV 27 (7 May 1959)
Comments: Stricken 1 Aug 1959

Hull No. and Name: 78 Savo Island
Date of Commission: 3 Feb 1944; Decommission or Loss***: 12 Dec 1946
Designations: CHVE 78 (12 Jun 1955); AKV 28 (7 May 1959)
Comments: Stricken 1 Sep 1959

Hull No. and Name: 79 Ommaney Bay
Date of Commission: 11 Feb 1944; Decommission or Loss***: 4 Jan 1945
Designations:
Comments: Lost to enemy action

Hull No. and Name: 80 Petrof Bay
Date of Commission: 18 Feb 1944; Decommission or Loss***: 30 Jul 1946
Designations: CVU 80 12 Jun 1955
Comments: Stricken 27 Jun 1958

Hull No. and Name: 81 Rudyerd Bay
Date of Commission: 25 Feb 1944; Decommission or Loss***: 11 Jun 1946
Designations: CVU 81 (12 Jun 1955); AKV 29 (7 May 1959)
Comments: Stricken 1 Aug 1959

Hull No. and Name: 82 Saginaw Bay
Date of Commission: 2 Mar 1944; Decommission or Loss***: 19 Jun 1946
Designations: CVHE 82 12 Jun 1955
Comments: Stricken 1 Mar 1959

Hull No. and Name: 83 Sargent Bay
Date of Commission: 9 Mar 1944; Decommission or Loss***: 23 Jul 1946
Designations: CVU 83 12 Jun 1955
Comments: Stricken 27 Jun 1958

Hull No. and Name: 84 Shamrock Bay
Date of Commission: 15 Mar 1944; Decommission or Loss***: 6 Jul 1946
Designations: CVU 84 12 Jun 1955
Comments: Stricken 27 Jun 1958

Hull No. and Name: 85 Shipley Bay
Date of Commission: 21 Mar 1944; Decommission or Loss***: 28 Jun 1946
Designations: CVHE 85 12 Jun 1955
Comments: Stricken 1 Mar 1959

Appendix III

The Evolution of Aircraft Carriers

Hull No. and Name: 86 Sitkoh Bay
Date of Commission: 28 Mar 1944; Decommission or Loss***: 27 Jul 1954
Designations: CVU 86 (12 Jun 1955); AKV 30 (7 May 1959)
Comments: Stricken 1 Apr 1960

Hull No. and Name: 87 Steamer Bay
Date of Commission: 4 Apr 1944; Decommission or Loss***: 8 Aug 1946
Designations: CVHE 87 12 Jun 1955
Comments: Stricken 1 Mar 1959

Hull No. and Name: 88 Cape Esperance
Date of Commission: 9 Apr 1944; Decommission or Loss***: 15 Jan 1959
Designations: CVU 88 12 Jun 1955
Comments: Stricken 1 Mar 1959

Hull No. and Name: 89 Takanis Bay
Date of Commission: 15 Apr 1944; Decommission or Loss***: 1 May 1946
Designations: CVU 89 (12 Jun 1955); AKV 31 (7 May 1959)
Comments: Stricken 1 Aug 1959

Hull No. and Name: 90 Thetis Bay
Date of Commission: 21 Apr 1944; Decommission or Loss***: 1 Mar 1964
Designations: CVHA 1 (1 Jul 1955); LPH 6 (28 May 1959)
Comments: Stricken 1 Mar 1964

Hull No. and Name: 91 Makassar Strait
Date of Commission: 27 Apr 1944; Decommission or Loss***: 9 Aug 1946
Designations: CVU 91 12 Jun 1955
Comments: Stricken 1 Sep 1958

Hull No. and Name: 92 Windham Bay
Date of Commission: 3 May 1944; Decommission or Loss***: Jan 1959
Designations: CVU 92 12 Jun 1955
Comments: Stricken 1 Feb 1959

Hull No. and Name: 93 Makin Island
Date of Commission: 9 May 1944; Decommission or Loss***: 19 Apr 1946
Designations:
Comments: Stricken 5 Jun 1946

Hull No. and Name: 94 Lunga Point
Date of Commission: 14 May 1944; Decommission or Loss***: 24 Oct 1946
Designations: CVU 94 (12 Jun 1955); AKV 32 (7 May 1959)
Comments: Stricken 1 Apr 1960

Hull No. and Name: 95 Bismarck Sea
Date of Commission: 20 May 1944; Decommission or Loss***: 21 Feb 1945
Designations:
Comments: Lost to enemy action

Hull No. and Name: 96 Salamaua
Date of Commission: 26 May 1944; Decommission or Loss***: 9 May 1946
Designations:
Comments: Stricken 21 May 1946

Appendix III

The Evolution of Aircraft Carriers

Hull No. and Name: 97 Hollandia
Date of Commission: 1 Jun 1944; Decommission or Loss***: 17 Jan 1947
Designations: CVU 97 (12 Jun 1955); AKV 33 (7 May 1959)
Comments: Stricken 1 Apr 1960

Hull No. and Name: 98 Kwajalein
Date of Commission: 7 Jun 1944; Decommission or Loss***: 16 Aug 1946
Designations: CVU 98 (12 Jun 1955); AKV 34 (7 May 1959)
Comments: Stricken 1 Apr 1960

Hull No. and Name: 99 Admiralty Islands
Date of Commission: 13 Jun 1944; Decommission or Loss***: 24 Apr 1946
Designations:
Comments: Stricken 8 May 1946

Hull No. and Name: 100 Bougainville
Date of Commission: 18 Jun 1944; Decommission or Loss***: 3 Nov 1946
Designations: CVU 100 (12 Jun 1955); AKV 35 (7 May 1959)
Comments: Stricken 1 Apr 1960

Hull No. and Name: 101 Matanikau
Date of Commission: 24 Jun 1944; Decommission or Loss***: 11 Oct 1946
Designations: CVU 101 (12 Jun 1955); AKV 36 (7 May 1959)
Comments: Stricken 1 Apr 1960

Hull No. and Name: 102 Attu
Date of Commission: 30 Jun 1944; Decommission or Loss***: 8 Jun 1946
Designations:
Comments: Stricken 3 Jul 1946

Hull No. and Name: 103 Roi
Date of Commission: 6 Jul 1944; Decommission or Loss***: 9 May 1946
Designations:
Comments: Stricken 21 May 1946

Hull No. and Name: 104 Munda
Date of Commission: 8 Jul 1944; Decommission or Loss***: 13 Sep 1946
Designations: CVU 104 12 Jun 1955
Comments: Stricken 1 Sep 1958

Hull No. and Name: 105 Commencement Bay
Date of Commission: 27 Nov 1944; Decommission or Loss***: 30 Nov 1946
Designations: CVHE 105 (12 Jun 1955); AKV 37 (7 May 1959)
Comments: Stricken 1 Apr 1971

Hull No. and Name: 106 Block Island
Date of Commission: 30 Dec 1944; Decommission or Loss***: 27 Aug 1954
Designations: LPH 1 (22 Dec 1957); CVE 106 (17 Feb 1959); AKV 38 (7 May 1959)
Comments: Stricken 1 Jul 1959

Hull No. and Name: 107 Gilbert Islands
Date of Commission: 5 Feb 1945; Decommission or Loss***: 15 Jan 1955
Designations: AKV 39 7 May 1959
Comments: Stricken 1 Jun 1961

Appendix III

The Evolution of Aircraft Carriers

Hull No. and Name: 108 Kula Gulf
Date of Commission: 12 May 1945; Decommission or Loss***: 15 Dec 1955
Designations: AKV 8 7 May 1959
Comments: Stricken 15 Sep 1970

Hull No. and Name: 109 Cape Gloucester
Date of Commission: 5 Mar 1945; Decommission or Loss***: 5 Nov 1946
Designations: CVHE 109 (12 Jun 1955); AKV 9 (7 May 1959)
Comments: Stricken 1 Apr 1971

Hull No. and Name: 110 Salerno Bay
Date of Commission: 19 May 1945; Decommission or Loss***: 16 Feb 1954
Designations: AKV 10 7 May 1959
Comments: Stricken 1 Jun 1960

Hull No. and Name: 111 Vella Gulf
Date of Commission: 9 Apr 1945; Decommission or Loss***: 9 Aug 1946
Designations: CVHE 111 (12 Jun 1955); AKV 11 (7 May 1959)
Comments: Stricken 1 Jun 1960

Hull No. and Name: 112 Siboney
Date of Commission: 14 May 1945; Decommission or Loss***: 31 Jul 1956
Designations: AKV 12 7 May 1959
Comments: Stricken 1 Jun 1970

Hull No. and Name: 113 Puget Sound
Date of Commission: 18 Jun 1945; Decommission or Loss***: 18 Oct 1946
Designations: CVHE 113 (12 Jun 1955); AKV 13 (7 May 1959)
Comments: Stricken 1 Jun 1960

Hull No. and Name: 114 Rendova
Date of Commission: 22 Oct 1945; Decommission or Loss***: 30 Jun 1955
Designations: AKV 14 7 May 1959
Comments: Stricken 1 Apr 1971

Hull No. and Name: 115 Bairoko
Date of Commission: 16 Jul 1945; Decommission or Loss***: 18 Feb 1955
Designations: AKV 15 7 May 1959
Comments: Stricken 1 Apr 1960

Hull No. and Name: 116 Badoeng Strait
Date of Commission: 14 Nov 1945; Decommission or Loss***: 17 May 1957
Designations: AKV 16 7 May 1959
Comments: Stricken 1 Dec 1970

Hull No. and Name: 117 Saidor
Date of Commission: 4 Sep 1945; Decommission or Loss***: 12 Sep 1947
Designations: CVHE 117 (12 Jun 1955); AKV 17 (7 May 1959)
Comments: Stricken 1 Dec 1970

Hull No. and Name: 118 Sicily
Date of Commission: 27 Feb 1946; Decommission or Loss***: 4 Oct 1954
Designations: AKV 18 7 May 1959
Comments: Stricken 1 Jul 1960

Appendix III

The Evolution of Aircraft Carriers

Hull No. and Name: 119 Point Cruz
Date of Commission: 16 Oct 1945; Decommission or Loss***: 31 Aug 1956
Designations: AKV 19 (7 May 1959)
Comments: Stricken 15 Sep 1970

Hull No. and Name: 120 Mindoro
Date of Commission: 4 Dec 1945; Decommission or Loss***: 4 Aug 1955
Designations: AKV 20 7 May 1959
Comments: Stricken 1 Dec 1959

Hull No. and Name: 121 Rabaul
Date of Commission:; Decommission or Loss***:
Designations: CVHE 121 (12 Jun 1955); AKV 21 (7 May 1959)
Comments: Stricken 1 Sep 1971. Inactivated after trials on 30 Aug1946, never commissioned

Hull No. and Name: 122 Palau
Date of Commission: 15 Jan 1946; Decommission or Loss***: 15 Jun 1954
Designations: AKV 22 7 May 1959
Comments: Stricken 1 Apr 1960

Hull No. and Name: 123 Tinian
Date of Commission:; Decommission or Loss***:
Designations: CVHE 123 (12 Jun 1955); AKV 23 (7 May 1959)
Comments: Stricken 1 Jun 1970. The ship was accepted by the Navy on 30 Jul 1946 but never commissioned.

* There were a number of carriers that were Decommissioned and then recommisioned for further service. Only the final Decommissioning date is listed for these carriers. Several carriers were also placed out of commission during major renovations or yard periods.

Note: In some cases the records regarding decommissioning dates were not known and incomplete. Consequently, the Decommissioning date in certain ships was left blank.